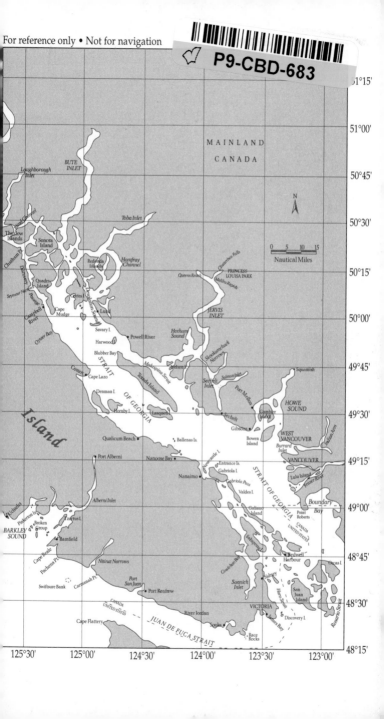

For reference only • Not for navigation

P9-CBD-683

MAINLAND
CANADA

N

0   5   10   15
Nautical Miles

Loughborough Inlet
BUTE INLET
Toba Inlet
Thurlow Islands
Sonora Island
Chatham
Discovery Passage
Redonda Islands
Homfray Channel
Quadra Island
Seymour Narrows
Cortes Island
Desolation Sound
Cape Mudge
Campbell River
Lund
Oyster Bay
Savary I.
Powell River
Harwood I.
Blubber Bay
Comox
Cape Lazo
Denman I.
Hornby I.
Qualicum Beach
Port Alberni
Nanoose Bay
Nanaimo

PRINCESS LOUISA PARK
Queens Reach
Malibu Rapids
Chatterbox Falls

JERVIS INLET

Hotham Sound

Skookumchuck Narrows

Salmon Inlet

Secret Inlet

Malaspina Strait
Texada Island
Lasqueti I.
Ballenas Is.
Jervis Inlet
Gibson
Sechelt Inlet
Sechelt
Gibsons
Bowen Island
Port Mellon
Squamish
HOWE SOUND
Gambier Island
WEST VANCOUVER
Burrard Inlet
VANCOUVER
Indian Arm

Island

STRAIT OF GEORGIA

Entrance Is.
Gabriola I.
Gabriola Pass
Valdes I.
Galiano Island
Saltspring I.
Point Roberts
Boundary Bay
CANADA UNITED STATES
Lulu Island
Fraser River

STRAIT OF GEORGIA

Ucluelet
Pipestem I.
BARKLEY SOUND
Broken Group
Effingham I.
Bamfield
Alberni Inlet
Cape Beale
Pachena Pt.
Swiftsure Bank
Carmanah Pt.
Nitinat Narrows
Port San Juan
Port Renfrew

CANADA UNITED STATES

Cape Flattery

JUAN DE FUCA STRAIT

River Jordan
Sooke
Bedwell Harbour
Cowichan Bay
Sidney
Saanich Inlet
San Juan Island
VICTORIA
Harro Strait
James Bay
Discovery I.
Race Rocks
Duncas I.
Rosario Strait

51°15'
51°00'
50°45'
50°30'
50°15'
50°00'
49°45'
49°30'
49°15'
49°00'
48°45'
48°30'
48°15'

125°30'   125°00'   124°30'   124°00'   123°30'   123°00'

GPS
WAYPOINTS

BRITISH COLUMBIA COAST

3000 WAYPOINTS FOR NAMED POSITIONS

HARBOUR AND INLET ENTRANCES

ANCHOR SITES • PUBLIC FLOATS

BUOYS • LIGHT HOUSES

COMPLETE WITH CHART NUMBER
AND HORIZONTAL DATUMS

BY DON DOUGLASS
FOREWORD BY KEVIN MONAHAN

1997/98 EDITION

Fine Edge
Productions

Important Legal Disclaimer

This waypoint list is for reference and planning purposes only. It is not to be used for navigation until and unless each waypoint has been validated by plotting on a nautical chart by a qualified user. Prior to using any waypoint listed, you, the user, must assume full responsibility for the accuracy and appropriateness of any waypoint, as well as applying correct safety margins and horizontal datum. Furthermore, this waypoint list may inadvertently contain typographical errors, or errors of content.

There is no actual or implied warranty as to the accuracy or suitability of these waypoints for any uses whatsoever. The author, publisher, and sources of cartographic information expressly disclaim any responsibility for any loss or damage resulting from the use of these waypoints or information.

The waypoints in this book were calculated by the author and verified from official government sources and are believed to be correct. Every effort (within limited resources) has been made to assure the accuracy and quality of this publication. Most of the waypoints in this book—widely published by the author and used by many skippers—have stood the test of time, and the few errors reported from these guidebooks have been corrected. Before you make a critical navigational decision you must consult official sources, including nautical charts and the monthly Notices to Mariners.

We invite you to send a stamped, self-addressed envelope to Fine Edge for free copies of a waypoint errata sheet. Or you can visit the Fine Edge web site (www.fineedge.com) to download errata sheets as they become available.

Front cover photo by Maria Steernberg • Back cover photo by Robin Hill-Ward
Design by Melanie Haage • Illustrations by Iain Lawrence

Library of Congress Cataloging-in-Publication Data

Douglass, Don, 1932–
    GPS waypoints British Columbia coast / by Don Douglass. -- 1997/98 ed.
      p.   cm.
    ISBN 0-938665-50-2
    1. Geographical positions--British Columbia. 2. Global
Positioning System.   I. Title.
G110.D68   1997
623.39'29711--d21
                                  97-20242
                                      CIP

ISBN 0-938665-50-2

PRINTED IN CANADA
ON 100% RECYCLED,
20% POST-CONSUMER PAPER

Foreword

Naturally beautiful British Columbia has a wonderful scenic coastline which can be enjoyed more when you know where you are and where you want to go. And now, the Global Positioning System (GPS) can help you find your own precise location to within a few boat lengths and guide you to the exact location of any waypoint you choose. To calculate the steering information you need, however, a GPS instrument requires you to input waypoints.

GPS Waypoints gives you many important waypoints, as well as prudent cautions to think about before you blindly program your unit. Before you accept someone else's waypoints, you have the responsibility to validate them for your own particular use.

This book, which contains the most complete listing of waypoint positions for the British Columbia coast yet published, is an important reference for all pleasure and work boats. Unique to this book is a listing for almost every anchorage along the B.C. coast, including hundreds of small unnamed coves, and by using it in conjunction with the listed charts, you can plan your next trip with ease and include many of these little-known places.

Based upon his years of cruising in British Columbia, Don Douglass has personally calculated most of the waypoints cited in this book, and because the waypoints listed are precise to within two decimal places, you can drop your anchor just about where he dropped his.

The praise Don's cruising guidebooks have received testifies to the care and integrity he teas taken in compiling this vital information, thus assuring you a safe and pleasant voyage. His popular cruising guides were spotlighted during a Microsoft presentation at the 1997 Seattle Boat Show as the first books to offer website up-dates of GPS waypoints. (http://www.fineedge.com)

This concise little reference should be found on every boat that heads into the waters of British Columbia.

—Kevin Monahan

Kevin Monahan, principal author of GPS Instant Navigation, works as a Ship's Officer for the Canadian Coast Guard. He lives in Victoria B.C.

How to Use This Book

The waypoints listed in this book are for reference and planning purposes to help you plan a trip and identify landmarks, navigation aids and seldom visited, unnamed coves. The alphabetical listings are convenient and easy to read. Because of the chance of groundings or collisions from misuse, these waypoints are not to be used for navigation until you have validated each one by plotting it correctly on a nautical chart. A well proven list of waypoints and a GPS unit can greatly add to your pleasure and safety when exploring the coast of British Columbia.

A waypoint is simply a precise "address" on a global grid. It is single piece of information to be treated with caution and skepticism until you have verified the accuracy of that waypoint, determined that you have an adequate margin of safety, and that it is safe to proceed to that point.

You can check the accuracy and appropriateness of any waypoint in this book by plotting its position on a nautical chart. You can modify these waypoints to fit your own judgment and needs so that, in effect, they become your waypoints.

A GPS steering diagram does not automatically indicate an intervening land mass or obstruction between the waypoint you have input and your present position; only by charting your waypoints and routes and by maintaining an alert lookout can you avoid these obstructions. If you were to program your GPS to take you directly to the lat/long of a lighthouse, for example, you might find yourself aground on the very reef or rock you wanted to avoid! Stories abound of groundings and collisions caused by skippers who blindly followed steering directions generated by a GPS receiver. And as experienced Vancouver sailor, June Cameron, noted recently in a letter to *Pacific Yachting:* "God looks after fools, but She has little patience with idiots!"

When you select a specific waypoint, be sure to take into account the following criteria: l) potential for error (waypoint, chart or in the GPS unit), 2) special local conditions—nearby hazards, currents, marginal weather, reduced visibility, traffic patterns, etc., 3) how the waypoint "fits" into a route (the direction from which you approach or leave that waypoint and the appropriate cross-track error limit).

To learn how to select waypoints, study a good text; otherwise you may find that "garbage in equals garbage out " Because of the inherent GPS error of 100 meters or more, be sure that any waypoint you use contains an adequate margin of safety for the environment and situation you anticipate. One rule of thumb is to add 0.25 mile to the safe side of that position (such as a buoy or navigation aid) before you insert it into a GPS receiver. In this book, we have not added a margin of safety for any of the waypoints. The nominal position of each buoys and nav-aid has been used. Only after you have plotted a waypoint (and analyzed a straight line drawn between waypoints) can you consider a waypoint or route to be safe. The Sand Heads light off the Frazer River is listed at 49°06.19' N, 123°18.57' W (NAD 83). This is a unique address, precise to about 60 feet for the nav-aid at the end of Steveston Jetty. NAD 83 refers to the horizontal datum, or reference point used by the cartographers. Chart datums vary so much around the world (and even inside British Columbia), that confusion frequently develops as you change charts or even editions of charts. Therefore a waypoint is incomplete unless its source datum is indicated as part of the position, and you should be careful to set your GPS unit to the same horizontal datum as the chart on which you plotted your course. (Each waypoint in this book lists the chart number and horizontal datum for its source chart.)

Charts for British Columbia are referenced to horizontal datums NAD 27 or NAD 83 and, in a few cases—such as found ;n the Queen Charlotte Islands, the original chart datum is unknown extra precaution is required. The difference between a waypoint taken off a chart NAD 27 or NAD 83 can be as much as 200 meters or more. (You can easily see this difference on the charts for Calvert Island Light which are printed in both horizontal datums). The Global Positioning System uses a horizontal datum known as WGS 84. For all practical purposes NAD 83 is the same as WGS 84 and therefore interchangeable. Stored waypoints may be adversely affected as you shift between different horizontal datums, so you need to learn how your unit works in this regard. (Consult your instruction manual and/or *GPS Instant Navigation* for detailed examples.)

GPS Waypoints contains a compilation of 3000 waypoints of key coastal sites for British Columbia personally calculated by the author from his own experience and verified from official government sources. Each waypoint is formatted as follows:

Kiltuish Inlet Chart 3743;
entrance: 53°24.40′ N, 128°31 .70′ W;
anchor (outside narrows): 53°23.82′ N, I 28°30.23 ′ W (NAD 27)

Each cove, destination, or nav-aid is identified by bold-face type. This book lists many small, unnamed coves, and bays for which we have tried to find and use local names. Where we could find no local name or reference to a name in either Canadian charts or in C.H.S. sources *(Sailing Directions),* we used one that seemed appropriate. Both local names or new names are shown in quotation marks.

The first chart listed after the place name is always the largest scale available and the one we used to determine the latitude and longitude of a place. Where additional charts are listed, they are smaller scale charts that cover the area

Information following the chart number(s) identifies: a general **position,** a specific **anchor** site, a mid-channel **entrance** point, a **buoy,** or a navigational **light.** Latitude and Longitude to the nearest one-hundredth of a minute are specified. (The term "position" is used for a general site or an anchor site that we have not personally checked or are unsure about.) Buoy and light positions are taken from the 1995 Canadian Coast Guard *Light List,* with seconds of arc converted to decimal minutes

The Lat/Long of an anchor site is just that; it is not the entrance to a cove. GPS receivers set to an anchor site—or to any other position given—will take you directly there, whether there is an intervening land mass or not.

The bulk of these waypoints are taken from the series *Exploring the British Columbia Coast* (see Appendix) which describe the entire B.C. coast in detail, including standard harbours, marinas, and B.C. Marine parks. Unique to this series of guidebooks are hundreds of previously undocumented coves and anchor sites not listed on charts or in *Sailing Directions* (their names set off by quotation marks in this book). Many of the listed anchor sites are complex and require experienced piloting for safe passage. You should refer to the diagrams and directions given in these guidebooks for safety cautions and additional detail.

Also included are the buoys, navigation lights, and lighthouses from the 1995 *List of Lights* of the CHS, including the Fraser River (and with the exception of freshwater lakes). Positions given in seconds in the *List of Lights* have been converted into decimal units to be compatible with the rest of the waypoints in this book. *(Caution:* a number of errors have been found in the *List of Lights.)* Please note that all buoys are subject to moving off their nominal position, and their position should be carefully verified. See Notice to Mariners for the latest corrections.

Developing your own list of proven waypoints will simplify and enhance your commercial or pleasure boating experience. The author and publisher invite your comments on these waypoints and your review of subsequent updates on our website.

Please read the legal disclaimer on the publisher's page. There are many navigational challenges along the coast British Columbia, and although GPS can help make any trip safer and more pleasurable, you need to have a knowledge of how it works, including its limitations and sources of errors. We suggest that you consult a reference text, such as *GPS Instant Navigation by* Kevin Monahan, before you commit to using GPS for navigation.

Aaltanhash Inlet Chart 3739;
entrance: 53°07.50′ N, 128°34.10′ W
(NAD 27)

Abrams Island Light Chart 3726;
position: (on isl.): 52°32.18′ N, 128°49.83′ W
(NAD unknown)

"Absalom Cove' Charts 3761, 3927;
entrance: 53°50.42′ N, 130°36.54′ W;
position: 53°51.18′ N, 130°37.25′ W (NAD 27)

Acland Islands Light Buoy U60
Chart 3478;
position: (W. of the islands):
48°48.62′ N, 123°22.90′ W (NAD 83)

Actaeon Sound Chart 3547;
entrance: 50°55.82′ N, 127°08.43′ W;
Skeene Bay position:
50°56.93′ N, 127°08.33′ W (NAD 83)

Active Pass Chart 3473;
Georgina Point light:
48°52.41′ N, 123°17.41′ W;
Helen Point light: 48°51.47′ N, 123°20.63′ W;
Enterprise Reef light:
48°50.71′ N, 123°20.81′ W (NAD 27);
Chart 3748; Ben Mohr Rock light:
48°51.63′ N, 123°23.38′ W (NAD 83)

Active Pass Light Chart 3473;
position: (on Georgina Pt, Mayne Isl.):
48°52.41′ N, 123°17.41′ W (NAD 27)

Ada Cove Chart 3785;
entrance: 52°03.85′ N, 128°03.70′ W;
anchor: 52°03.51′ N, 128°03.09′ W (NAD 27)

Adams Harbour Chart 3784;
anchor: 51°41.15′ N, 128°06.17′ W (NAD 27)

Addenbroke Island Light Chart 3921;
position: (W. pt of isl.):
51°36.20′ N, 127°51.83′ W (NAD 83)

Addenbroke Light Station
Charts 3921, 3934;
position (light): 51°36.20′ N, 127°51.82′ W;
anchor: 51°36.35′ N, 127°51.80′ W (NAD 83)

"Addenbroke Point Cove" Chart 3934;
entrance: 51°31.50′ N, 127°47.50′ W;
anchor: 51°31.58′ N, 127°45.69′ W (NAD 83)

Admiral Group Chart 3786;
position (east turn at Admiral Group):
52°01.30′ N, 128°15.20′ W (NAD 27)

Adventure Cove Chart 3649;
anchor: 49°12.17′ N, 125°51.14′ W

**Agnew Bank Light and Bell Buoy
D24** Chart 3955;
position: (Porpoise Channel entrance):
54°11.45′ N, 130°20.34′ W (NAD 27)

Agnew Passage Light Chart 3514;
position: (NE shore of small islet):
49°46.47′ N, 123°59.28′ W (NAD 83)

Aguilar Point Sector Light Chart 3646;
position: (on the pt):
48°50.37′ N, 125°08.37′ W (NAD 83)

Ahclakerho Channel Chart 3931;
east entrance 51°17.48′ N, 127°23.28′ W
(NAD 83)

Ahlstrom Point Light Chart 3514;
position: (N. shore of Jervis Inlet):
49°46.80′ N, 124°08.53′ W (NAD 83)

Ahousat Chart 3643:
Ahousat fuel float: 49°16.94′ N, 126 04.20′ W
(NAD unknown)

Ala Narrows Chart 3746 (inset);
position: 53°31.35′ N, 129°53.51′ W (NAD 27)

Ala Passage North Chart 3746;
north entrance: 53°32.96′ N, 129°56.78′ W
(NAD 27)

Ala Passage South
Charts 3721, 3741, 3746;
fairway position (west of Peck Shoal):
53°27.86' N, 129°55.87' W; (NAD 27)

Alarm Cove Charts 3785, 3787;
entrance: 52°07.10' N, 128°07.10' W;
position: 52°07.04' N, 128°06.83' W (NAD 27)

Alarm Rock Light Chart 3477;
position: (on SE end of rock):
48°57.53' N, 123°40.45' W (NAD 27)

Alaska Ferry North Light Chart 3958;
position: 54°17.82' N, 130°21.11' W
(NAD 83)

Alaska Ferry South Light Chart 3958;
position: (on S. cradle dolphin):
54°17.77' N, 130°21.10' W (NAD 83)

Alberni Inlet Chart 3668;
entrance: 48°57' N, 125°01' W (NAD 83)

Alberni Light Chart 3668;
position: (W. side of channel, entrance to
Somass R.): 49°14.29' N, 124°49.34' W
(NAD 83)

Albert Head Light Chart 3419;
position: (SE part of Albert Head):
48°23.23' N, 123°28.60' W (NAD 83)

Alert Bay Breakwater Light
Chart 3546;
position: (NW extremity of breakwater):
50°35.39' N, 126°55.98' W (NAD 83)

Alert Bay Chart 3546 (inset);
breakwater light:
59°35.40' N, 126°55.88' W (NAD 83)

Alert Rock Light Buoy N17
Chart 3546;
position: (N. of rock):
50°34.97' N, 126°57.66' W (NAD 83)

Alexander Inlet Chart 3734;
entrance: 52°40.40' N, 128°34.50' W;
anchor: 52°38.33' N, 128°40.36' W (NAD 27)

**Alexandra Bank Bifurcation Light
Buoy DAX** Chart 3957;
position: 54°14.15' N, 130°33.95' W
(NAD 83)

Alexandra Passage Chart 3934;
south entrance (0.35 mile south Egg Rocks):
51°14.00' N, 127°50.00' W (NAD 83)

Alfred Island Chart 3538;
position: 50°11.88' N, 124°47.33' W
(NAD 27)

Alice Arm Chart 3920;
entrance Liddle Channel:
55°23.60' N, 129°41.50' W (NAD 83)

Alice Arm Light Chart 3920;
position: (S. of Hans Pt):
55°25.56' N, 129°40.07' W (NAD 83)

Alice Arm Settlement Chart 3920;
wharf position: 55°28.28' N, 129°29.72' W
(NAD 83)

Alice Arm Wharf Light Chart 3920;
position: (on wharf):
55°28.26' N, 129°29.74' W (NAD 83)

Alison Sound Chart 3552;
entrance: 51°07.55' N, 127°07.80' W (NAD 27)

**Allan Rocks Light and
Whistle Buoy N33** Chart 3549;
position: 51°01.60' N, 127°37.53' W (NAD 83)

Allard Bay Chart 3931;
entrance: 51°26.40' N, 127°19.20' W;
position: 51°27.75' N, 127°19.07' W (NAD 83)

"Allcroft Point Cove" Chart 3746;
entrance: 53°35.65' N, 130°03.40' W;
anchor: 53°35.76' N, 130°02.98' W (NAD 27)

Allen Point Light Chart 3745;
position: (on pt S. side of channel, Europa
Reach): 53°26.53' N, 128°24.18' W (NAD 27)

Allies Island Chart 3541;
anchor: 50°12.80' N, 124°48.85' W (NAD 27)

Alliford Bay Chart 3890 (inset);
entrance: 53°12.90' N, 131°59.60' W;
anchor: 52°12.62' N, 131°59.21' W (NAD 27)

Alliford Bay Light Chart 3890;
position: (on isl., SW of Kwuna Pt):
53°12.80' N, 131°59.51' W (NAD 27)

Allison Harbour Chart 3921;
entrance: 51°02.15' N, 127°31.23' W;
anchor head of bay:
51°03.47' N, 127°30.38' W;
anchor east cove:
51°02.94' N, 127°30.50' W (NAD 83)

Alma Russell Islands Chart 3671;
anchor: 48°57.09' N, 125°12.23' W (NAD 27)

Alpha Bay Charts 3773, 3927;
position: 53°51.95' N, 130°17.10' W (NAD 27)

Alston Cove Chart 3737;
entrance: 52°45.10' N, 128°45.60' W;
anchor: 52°44.83' N, 128°44.58' W (NAD 27)

"Althorp Point Cove" Chart 3544;
50°28.38' N, 125°47.72' W (NAD 83)

Amai Inlet Chart 3682
entrance: 50°01.00' N, 127°10.00' W;
(NAD 27)

Amos Island Light Chart 3683;
position: 50°00.78' N, 127°21.08' W
(NAD 27)

Amphitrite Point Light Chart 3646;
position: (extremity of pt, entrance to
Ucluelet Hbr): 48°55.28' N, 125°32.38' W
(NAD 83)

Anchor Bight Charts 3934, 3931;
anchor: 51°16.75' N, 127°38.88' W (NAD 83)

Anchor Cove Chart 3891;
anchor: 53°12.38' N, 132°14.67' W (NAD 83)

Anchor Cove Chart 3931;
anchor: 51°17.23' N, 127°22.68' W (NAD 83)

"Anchor Islands Cove" Chart 3550;
entrance: 51°05.80' N, 127°33.30' W;
anchor: 51°05.97' N, 127°33.11' W (NAD 83)

Anchorage Cove Chart 3515;
position: 50°54.40' N, 126°12.00' W
(NAD 83)

Anderson Bay Charts 3512, 3311;
anchor: 49°31.03' N, 124°08.08' W (NAD 27)

Anderson Cove Chart 3641;
anchor: 48°21.72' N, 123°39.29' W (NAD 27)

Anderson Passage Chart 3719 (inset);
entrance: 53°07.53' N, 129°33.26' W;
anchor (southeast nook):
53°07.94' N, 129°31.81' W (NAD 27)

Anderson Point Light Chart 3664;
position: 49°38.79' N, 126°28.20' W
(NAD 27)

Anger Inlet Chart 3746;
entrance: 53°31.35' N, 129°58.40' W (NAD 27)

Angler Cove Charts 3740 or 3742;
position: 53°18.90' N, 128°53.10' W (NAD 27)

Anna Inlet Chart 3807;
entrance: 52°42.98' N, 131°49.60' W;
anchor: 52°42.22' N, 131°50.43' W (NAD 27)

Annacis Island Pile Wall Centre Light
Chart 3490;
position: (inside pile wall):
49°11.14' N, 122°55.24' W (NAD 83)

Annacis Island Pile Wall South Light
Chart 3490;
position: (S. end, pile wall):
49°10.36′ N, 122°55.59′ W (NAD 83)

Annette Inlet Chart 3313, p. 10;
anchor: 49°49.63′ N, 123°23.61′ W (NAD 83)

Annie's Inlet Charts 3747, 3927;
entrance: 53°44.80′ N, 130°18.30′ W;
anchor: 53°44.51′ N, 130°18.23′ W (NAD 27)

**Annieville Channel Pile Wall
North Light** Chart 3490;
position: (N. end of pile wall):
49°10.96′ N, 122°55.02′ W (NAD 83)

**Annieville Channel Pile Wall
South Light** Chart 3490;
position: (S. end of pile wall):
49°10.83′ N, 122°55.09′ W (NAD 83)

Annieville Channel Range Light
Chart 3490;
position: (S. bank of river):
49°09.89′ N, 122°55.76′ W (NAD 83)

Annieville Rock Wall 1 Light
Chart 3490;
position: (S. side): 49°11.18′ N, 122°54.90′ W
(NAD 83)

Annieville Rock Wall 3 Light
Chart 3490;
position: (opposite Shoal Pt):
49°11.62′ N, 122°54.62′ W (NAD 83)

Annieville Rock Wall 4 Light
Chart 3490;
position: (NE of Shoal Pt):
49°11.80′ N, 122°54.41′ W (NAD 83)

Annieville Rock Wall 5 Light
Chart 3490;
position: (upper end of wall):
49°12.00′ N, 122°53.99′ W (NAD 83)

Anthony Island Cove Chart 3825;
entrance: 52°05.57′ N, 131°12.28′ W;
anchor (small east bight):
52°05.77′ N, 131°12.76′ W (NAD 27)
**Anthony Island Provincial Park, Sgan
Gwaii World Heritage Site** Chart 3825

Anvil Island Chart 3526;
north bight: 49°31.65′ N, 123°17.36′ W
(NAD 27);
south bight: 49°31.28′ N, 123°17.33′ W
(NAD 27)

Anyox Chart 3920;
position: 55°25.10′ N, 129°48.73′ W (NAD 83)

Apodaca Cove Chart 3526;
position: 49°21.15′ N, 123°20.03′ W (NAD 27)

Apple Bay Chart 3679;
anchor: 50°36.11′ N, 127°39.30′ W (NAD 83)

"April Point Cove" Charts 3540, 3539;
anchor: 50°03.72′ N, 125°13.50′ W (NAD 83)

Arachne Reef Light Chart 3441;
position: (N. side of reef):
48°41.10′ N, 123°17.54′ W (NAD 27)

Archibald Islands Light Chart 3957;
position: (northernmost rocky islet of
Archibald Isl. group):
54°12.89′ N, 130°50.05′ W (NAD 83)

Argyh Cove Chart 3737;
position: 52°54.48′ N, 129°01.85′ W (NAD 27)

Armentieres Channel
Chart 3891 (inset);
north entrance: 53°07.70′ N, 132°23.90′ W;
anchor (buoy): 53°06.65′ N, 132°23.50′ W
(NAD 83)

Arran Rapids Chart 3543;
west entrance: 50°25.10′ N, 125°08.60′ W
(NAD 27)

Arrow Passage Chart 3515;
west entrance: 50°42.10′ N, 126°42.25′ W
(NAD 83)

Arthur Passage Chart 3773;
south entrance: 53°57.80′ N, 130°11.90′ W;
north entrance: 54°04.50′ N, 130°16.80′ W;
Herbert Reefs: 54°01.40′ N, 130°14.23′ W
(NAD 27)

Ashdown Island Light Chart 3742;
position: (on westerly of isl.):
53°03.72′ N, 129°13.74′ W (NAD 27)

Assits Island Light Chart 3668;
position: 48°56.27′ N, 125°01.94′ W
(NAD 83)

Atkins Bay Chart 3743;
position: 53°50.90′ N, 128°33.60′ W
(NAD 27)

Atkins Cove and "Early Bird Cove"
Chart 3679;
anchor Atkins: 50°30.57′ N, 127°34.70′ W;
anchor Early Bird: 50°30.85′ N, 127°34.83′ W
(NAD 83)

Atrevida Point Light Chart 3664;
position: (on the pt):
49°39.20′ N, 126°26.38′ W (NAD 27)

Atrevida ReefLight Buoy Q26
Chart 3311;
position: 49°55.00′ N, 124°40.00′ W
(NAD 83)

Attwood Bay Charts 3541, 3312;
position: "North Cove":
50°19.03′ N, 124°39.96′ W (NAD 27)

Bacchante Bay Chart 3674;
anchor: 49°27.15′ N, 126°02.25′ W (NAD 83)

Bachelor Bay Chart 3730;
position: 52°22.05′ N, 126°54.60′ W (NAD 27)

Baeria Rocks Light Chart 3671;
position: 48°56.99′ N, 125°09.22′ W
(NAD 27)

Bag Harbour Chart 3809;
entrance: 52°20.88′ N, 131°20.96′ W;
anchor: 52°20.80′ N, 131°21.80′ W (NAD 27)

"Baidarka Cove" Chart 3683;
anchor 50°09.60′ N, 127°39.35′ W (NAD 27)

Bainbridge Cove Chart 3720;
entrance: 52°11.70′ N, 127°59.00′ W (NAD 27)

**Bajo Reef Light and
Whistle Buoy M56** Chart 3662;
position: (S. of Bajo Reef):
49°33.80′ N, 126°50.00′ W (NAD 27)

Baker Inlet Charts 3772 (inset), 3773;
anchor (south bight):
53°48.98′ N, 129°56.65′ W;
anchor (inlet head):
53°48.52′ N, 129 51.08′ W (NAD 27)

Baker Inlet Light Chart 3772;
position: (marking entrance):
53°48.69′ N, 129°57.13′ W (NAD 27)

Baker Point Chart 3737; light:
52°48.20′ N, 129°12.82′ W (NAD 27)

Baker Point Light Chart 3737;
position: (on pt): 52°48.20′ N, 129°12.82′ W
(NAD 27)

Ballenas Islands Chart 3512;
north island light:
49°21.03′ N, 124°09.53′ W (NAD 27)

Ballenas Islands Light Chart 3512;
position: (N. pt of N. Ballenas Isl.):
49°21.03′ N, 124°09.53′ W (NAD 27)

Ballet Bay Chart 3312;
anchor: 49°43.00′ N, 124°10.77′ W (NAD 27)

Bamber Point Light Chart 3525;
position: (on the pt):
50°41.73′ N, 126°13.93′ W (NAD unknown)

Bamfield Inlet Chart 3646;
anchor: 48°49.83′ N, 125°08.33′ W (NAD 83)

Bamford Lagoon Chart 3552;
entrance: 50°59.90′ N, 127°15.10′ W;
anchor: 50°59.40′ N, 127°18.25′ W (NAD 27)

Barber Passage Chart 3543 (inset);
center passage position:
50°23.75′ N, 125°09.00′ W (NAD 27)

Bare Point Light Chart 3475;
position: (extremity of pt):
48°55.78′ N, 123°42.28′ W (NAD 27)

Bargain Bay Chart 3535 (inset);
anchor: 49°36.90′ N, 124°02.08′ W (NAD 27)

Bark Island Light Chart 3785;
position: (E. side of isl.):
52°10.02′ N, 128°02.87′ W (NAD 27)

Barnard Harbour Chart 3723 (inset);
Aikman Passage entrance:
53°04.77′ N, 129°07.38′ W;
Burnes Passage entrance:
53°05.08′ N, 129°06.60′ W;
anchor (extreme east):
53°04.16′ N, 129°05.76′ W (NAD 27)

Barnes Bay Chart 3537 (inset);
islet position: 50°19.45′ N, 125°15.30′ W
(NAD 27)

Baronet Passage Charts 3545, 3546;
east end: 50°34.05′ N, 126°30.35′ W (NAD 83)

Barrett Rock Light Chart 3958;
position: (on shore of reef):
54°14.58′ N, 130°20.54′ W (NAD 83)

Barter Cove Charts 3651, 3682, 3683;
anchor: 50°00.8′ N, 127°23.2′ W (NAD 83)

Base Sand Light Buoy D10 Chart 3717;
position: 54°05.20′ N, 130°15.63′ W (NAD 83)

"Baseball Bay" Chart 3648;
anchor: 49°22.48′ N, 126°13.60′ W (NAD 27)

Bate Point Light Buoy P12 Chart 3457;
position: 49°10.55′ N, 123°56.03′ W (NAD 27)

Battle Bay Chart 3683, 3680;
anchor: 50°06.92′ N, 127°35.50′ W (NAD 27)

Bauza Cove Chart 3546;
anchor: 50°32.62′ N, 126°49.13′ W (NAD 83)

Bawden Bay Chart 3648;
inner north nook anchor:
49°17.03′ N, 126°00.35′ W (NAD 27)

Bay of Plenty Chart 3737;
entrance: 52°49.90′ N, 128°45.00′ W;
anchor: 52°50.50′ N, 128°46.10′ W (NAD 27)

Baynes Channel North Light
Chart 3424;
position: (SSE of Cadboro Pt):
48°27.02′ N, 123°15.83′ W (NAD 83)

Baynes Sound Light Chart 3527;
position: 49°28.35′ N, 124°41.03′ W (NAD 27)

Bazett Island Chart 3670;
anchor: 49°01.05′ N, 125°17.73′ W (NAD 83)

Bear Cove Chart 3548 (inset);
position: 50°43.45′ N, 127°27.55′ W (NAD 83)

Bear Point Light Chart 3544;
position: 50°21.93′ N, 125°40.30′ W (NAD 83)

Beattie Anchorage Chart 3894;
entrance: 53°01.55′ N, 131°54.35′ W;
anchor: 53°01.33′ N, 131°54.25′ W (NAD 27)

Beaumont Island Light Chart 3729;
position: (N. extremity of isl., Johnson
Channel): 52°17.72′ N, 127°56.60′ W (NAD 27)

"Beaumont Island" Chart 3720;
entrance (east): 52°17.45' N, 127°56.40' W;
anchor (south end):
52°17.40' N, 127°56.68' W (NAD 27)

Beaver Cove Chart 3546;
position: 50°32.50' N, 126°52.65' W
(NAD 83)

Beaver Harbour Chart 3548;
south entrance: 50°42.30' N, 127°22.70' W;
Round Island light:
50°43.58' N, 127°21.83' W (NAD 83)

Beaver Inlet Chart 3555 (inset);
anchor: 50°29.73' N, 125°37.97' W (NAD 27)

Beaver Passage Charts 3747, 3927;
west entrance (0.25 mile northwest of
Hankin Rock): 53°42.60' N, 130°25.00' W;
mid-passage turn (0.22 mile east of Connis
light): 53°45.48' N, 130°18.60' W;
north entrance (mid-channel Bully Island and
Kitkatla Island): 53°47.85' N, 130°20.00' W
(NAD 27)

Beaver Point Light Chart 3441;
position: (On N. tip of pt):
48°46.27' N, 123°21.97' W (NAD 27)

Becher Bay Chart 3641
entrance: 48°19.00' N, 123°37.10' W
(NAD 27)

Bedwell Bay Chart 3495;
anchor: 49°18.96' N, 122°55.05' W (NAD 83)

**Bedwell Harbour, South Pender
Island** Chart 3313, p. 11;
customs float: 48°44.83' N, 123°13.69' W
(NAD 83)

Bedwell Islands Light Chart 3679;
position: (S. end of islet):
50°28.55' N, 127°53.75' W (NAD 83)

Bedwell Sound Chart 3649
entrance: 49°15.00' N, 125°50.00' W (NAD 27)

Belize Inlet Chart 3552 (inset);
entrance: 51°07.70' N, 127°33.10' W (NAD 27)

Bella Bella Charts 3785, 3787; fuel dock:
52°09.77' N, 128°08.37' W (NAD 27)

Bella Coola Breakwater Light
Chart 3730;
position: 52°22.55' N, 126°47.70' W (NAD 27)

Bella Coola Chart 3730 (inset);
harbour entrance:
52°22.60' N, 126°47.60' W (NAD 27)

Belle Bay Chart 3933;
position: 55°17.45' N, 129°56.70' W (NAD 27)

Belleisle Sound Chart 3515;
entrance: 50°54.50' N, 126°25.30' W (NAD 83)

**Ben Mohr Rock Bifurcation Light
Buoy UK** Chart 3478;
position: (NE of Peile Point light):
48°51.63' N, 123°23.38' W (NAD 83)

Bennett Bay Charts 3313, p. 12;
anchor: 48°50.72' N, 123°14.84' W (NAD 83)

Benson Island Light Chart 3670;
position: (on N. extremity of Isl.):
48°53.16' N, 125°22.64' W (NAD 83)

Bent Harbour Chart 3710 (inset);
entrance: 52°30.18' N, 129°02.57' W;
anchor: 52°30.89' N, 129°03.13' W (NAD 27)

Berens Island Light Chart 3415;
position: (SE extremity of isl., W. of hbr
entrance): 48°25.45' N, 123°23.50' W
(NAD 27)

Beresford Inlet Chart 3808;
entrance: 52°35.35' N, 131°34.10' W;
anchor: 52°38.03' N, 131°37.10' W (NAD 27)

Bergh Cove Chart 3681;
entrance: 50°32.23' N, 127°37.55' W
(NAD 83)

Berry Inlet Chart 3728;
entrance: 52°15.85' N, 128°20.25' W;
anchor: 52°17.08' N, 128°17.78' W (NAD 27)

Berry Point Light Chart 3494;
position: (N. extremity of pt):
49°17.71' N, 122°59.22' W (NAD 83)

Berryman Cove Chart 3649;
entrance: 49°09.10' N, 125°40.00' W
(NAD 27)

Berryman Point Light Chart 3649;
position: (NW of pt):
49°09.38' N, 125°40.64' W (NAD 27)

Bessborough Bay Chart 3544;
anchor: 50°28.98' N, 125°45.93' W (NAD 83)

Best Point Light Chart 3495;
position: (on pt): 49°22.83' N, 122°52.78' W
(NAD 83)

Betteridge Inlet
Charts 3719 (inset), 3742;
entrance(Hale Rock):
53°04.93' N, 129°29.92' W;
anchor: 53°06.18' N, 129°30.12' W (NAD 27)

Beware Cove Chart 3545;
anchor: 50°35.83' N, 126°32.54' W (NAD 83)

Beware Passage Chart 3545;
50°34.43' N, 126°29.50' W (NAD 83)

Bickley Bay Chart 3543;
anchor: 50°26.77' N, 125°23.24' W (NAD 27)

Big Bay Chart 3543 (inset);
public float: 50°23.54' N, 125°08.11' W
(NAD 27)

Big Bay Charts 3959, 3963;
entrance(south passage):
54°27.45' N, 130°30.00' W;
anchor (Salmon Bight):
54°27.50' N, 130°25.00' W (NAD 83)

Big Frypan Bay Chart 3934;
entrance: 51°29.17' N, 127°42.36' W;
anchor south cove:
51°28.87' N, 127°42.82' W;
anchor 10-foot hole:
51°28.90' N, 127°43.16' W (NAD 83)

Bigsby Inlet, Darwin Sound
Chart 3808;
entrance: 52°36.90' N, 131°41.50' W
(NAD 27)

Billard Rock Light Buoy M3
Chart 3686;
position: (S. side of rock):
50°25.79' N, 127°57.80' W (NAD 83)

Billton Point Light Chart 3668;
position: (on pt): 49°00.74' N, 124°52.34' W
(NAD 83)

Billy Bay Charts 3761, 3927;
entrance: 53°49.83' N, 130°25.80' W;
anchor: 53°50.43' N, 130°25.97' W (NAD 27)

Billygoat Bay, Helmcken Island
Chart 3544;
sector light: 50°23.93' N, 125°51.43' W;
anchor Billygoat Bay:
50°23.93' N, 125°52.00' W (NAD 83)

Bird Islet Light Chart 3481;
position: 49°21.77' N, 123°17.43' W
(NAD 27)

Birnie Island Light Chart 3963;
position: (on rock, S. of Knox Pt):
54°35.42' N, 130°27.72' W (NAD 83)

Bishop Bay, Hot Springs
Charts 3742, 3743;
entrance: 53°26.25' N, 128°54.50' W;
float: 53°28.20' N, 128°50.10' W (NAD 27)

"Bitter End Cove" Chart 3921;
anchor: 51°36.17' N, 127°40.40' W (NAD 83)

"Bitter End of Roscoe Inlet"
Chart 3940;
position: 52°27.02' N, 127°42.94' W
(NAD 83)

Blackney Passage Chart 3546;
south entrance: 50°33.10' N, 126°41.40' W
(NAD 83)

Blackrock Point Light Chart 3742;
position: 53°12.47' N, 129°20.55' W
(NAD 27)

Blair Inlet Charts 3710 (inset), 3728;
Ivory Island (light station):
52°16.18' N, 128°24.30' W (NAD 27)

Blenkinsop Bay Charts 3564, 3544;
anchor: 50°29.20' N, 126°00.57' W (NAD 83)

"Bligh Cove" Chart 3664;
anchor: 49°39.04' N, 126°31.15' W (NAD 27)

Blind Bay Chart 3312;
entrance: 49°43.10' N, 124°12.70' W
(NAD 27)

Blind Channel Chart 3543;
public float: 50°24.82' N, 125°30.00' W
(NAD 27)

Blinkhorn Peninsula Light Chart 3546;
position: (on peninsula):
50°32.62' N, 126°46.95' W (NAD 83)

Block Head Light Chart 3742;
position: (SE endof Farrant Isl.):
53°18.15' N, 129°20.00' W (NAD 27)

Block Islands Light Chart 3742;
position: (on small isl.):
53°08.98' N, 129°43.92' W (NAD 27)

Blow Reef Light Chart 3785;
position: (near entrance, Gunboat Passage
and Bella Bella): 52°10.90' N, 128°03.63' W
(NAD 27)

Blowhole Bay Chart 3663;
entrance: 49°49.70' N, 126°40.3' W (NAD 27)

Bloxam Passage Chart 3927;
east entrance: 54°01.83' N, 130°15.13' W;
west entrance: 54°01.33' N, 130°16.12' W
(NAD 27)

Blubber Bay Chart 3513;
anchor southeast corner:
49°47.65' N, 124°36.80' W (NAD 27)

Blue Heron Basin Entrance Light
Chart 3476;
position: 48°40.40' N, 123°24.67' W
(NAD 27)

Blue Heron Basin Light Chart 3476;
position: (boat basin entrance):
48°40.42' N, 123°25.05' W (NAD 27)

Blunden Bay Chart 3550;
position: 51°11.05' N, 127°46.70' W (NAD 83)

Blunden Harbour Chart 3548 (inset);
entrance: 50°54.07' N, 126°16.00' W;
anchor: 50°54.42' N, 127°17.37' W (NAD 83)

Blunden Harbour Chart 3548 (inset);
anchor: 50°54.42' N, 127°17.37' W (NAD 83)

Boat Bay Chart 3545;
anchor: 50°31.38' N, 126°33.84' W (NAD 83)

Boat Bay Light Chart 3545;
position: (W. of bay):
50°31.18' N, 126°34.62' W (NAD 83)

Boat Bluff Sector Light Chart 3711;
position: (western side of bluff, Sarah Isl.):
52°38.61' N, 128°31.35' W (NAD 27)

Boat Cove Chart 3512;
anchor: 49 27.81'N, 124 14.3'W (NAD 27)

Boat Harbour (Kenary Cove)
Chart 3313, p. 18;
position: 49°05.55' N, 123°47.90' W (NAD 83)

Boat Inlet Charts 3710 (inset), 3728;
entrance: 52°18.57' N, 128°21.70' W;
anchor: 52°18.51' N, 128°22.24' W (NAD 27)

Boat Nook, North Pender Island
Chart 3313, p. 11;
position: 48°45.95' N, 123°18.17' W (NAD 83)

Bob Bay Chart 3787;
entrance: 52°03.90' N, 128°06.90' W;
anchor: 52°02.24' N, 128°06.39' W (NAD 27)

Boddy Narrows Chart 3787;
south entrance: 52°07.20' N, 128°18.75' W
(NAD 27)

Bodega Anchorage Chart 3537;
anchor: 50°17.16' N, 125°13.40' W (NAD 27)

"Bodega Cove" Chart 3664;
anchor: 49°44.10' N, 126°38.19' W (NAD 27)

Bodega Light Chart 3664;
position: (NE of Bodega Isl.):
49°44.23' N, 126°37.34' W (NAD 27)

Boho Bay Charts 3312, 3512;
anchor: 49°29.85' N, 124°13.71' W (NAD 27)

Bolin Bay Chart 3962;
entrance: 52°50.25' N, 128°11.90' W;
anchor: 52°50.13' N, 128°12.93' W (NAD 27)

Bond Lagoon Chart 3547;
entrance: 50°56.80; N, 127°06.00' W (NAD 83)

Bone Anchorage Chart 3737;
anchor: 52°50.49' N, 129°00.60' W (NAD 27)

Bones Bay Chart 3545;
position: 50°35.12' N, 126°21.35' W (NAD 83)

Bonilla Island Chart 3747;
position (Bonilla Island light):
53°29.55' N, 130°38.10' W (NAD 27)

Bonilla Island Sector Light Chart 3747;
position: (W. side of isl.):
53°29.57' N, 130°38.15' W (NAD 27)

Bonilla Point Chart 3606;
light: 48°35.74' N, 124°42.98' W (NAD 27)

Bonilla Point Fisheries Light
Chart 3606;
position: (on pt): 48°35.74' N, 124°42.98' W
(NAD 27)

Booker Lagoon Chart 3547;
Booker Passage entrance:
50°46.65' N, 126°44.52' W (NAD 83)

Boot Cove, Saturna Island
Chart 3313, p. 12;
anchor: 48°47.52' N, 123°11.90' W (NAD 83)

"Bootleg Cove" Chart 3546;
50°42.80' N, 126°34.32' W (NAD 83)

Borde Island Light Chart 3742;
position: (northern pt of isl.):
53°05.14' N, 129°07.16' W (NAD 27)

Bordelais Islets Light Chart 3671;
position: (on largest islet):
48°49.08' N, 125°13.81' W (NAD 27)

Borrowman Bay, Turtish Harbour
Charts 3723 (inset), 3724;
entrance (Morison Passage):
52°44.56' N, 129°18.30' W (NAD 27)

Bosquet Point Light Chart 3934;
position: (extremity of Penrose Isl.):
51°30.62′ N, 127°44.03′ W (NAD 83)

Boston Point Light Chart 3664;
position: (on pt): 49°39.70′ N, 126°36.68′ W
(NAD 27)

"Boswell Cove" Chart 3931;
anchor: 51°22.53′ N, 127°26.47′ W (NAD 83)

Boswell Inlet Chart 3931;
entrance: 51°20.00′ N, 127°34.70′ W (NAD 83)

Bottleneck Cove ("Coyote Cove")
Chart 3648;
anchor: 49°26.77′ N, 126°12.90′ W (NAD 27)

Bottleneck Inlet Chart 3734;
entrance: 52°42.80′ N, 128°25.50′ W;
anchor: 52°42.58′ N, 128°24.05′ W (NAD 27)

Boughey Bay Chart 3545, 3564;
anchor: 50°31.18′ N, 126°11.10′ W (NAD 83)

Boukind Bay Chart 3940;
entrance: 52°27.00′ N, 127°56.30′ W;
anchor: 52°27.75′ N, 127°56.25′ W (NAD 83)

Boulton Bay ("Melibe Anchorage"),
Sheer Point Chart 3539;
anchor: 50°12.01′ N, 125°07.50′ W (NAD 83)

Boundary Bay
Chart 3463; flashing red light,
entrance to Nicomekl River:
49°00.75′ N, 122°56.18′ W (NAD 27)

Brackman Island Ecological Reserve
Chart 3313 (inset), p. 9;
anchor: 48°43.19′ N, 123°23.03′ W (NAD 83)

Brant Bay Chart 3742;
entrance: 53°15.30′ N, 129°22.42′ W (NAD 27)

Breakwater Island Light Chart 3475;
position: (W. side of isl.):
49°07.85′ N, 123°40.93′ W (NAD 27)

Brem Bay Chart 3541;
position west shore:
50°25.9′ N, 124°40.27′ W (NAD 27)

Bremner Bay Chart 3784;
entrance: 51°50.87′ N, 128°07.14′ W,
anchor: 51°50.85′ N, 128°06.75′ W (NAD 27)

Brigade Bay Chart 3526;
anchor: 49°29.42′ N, 123°20.00′ W (NAD 27)

"Briggs First Narrows" Chart 3940;
position: 52°21.68′ N, 128°00.61′ W (NAD 83)

Briggs Inlet Chart 3940;
entrance: 52°19.15′ N, 128°00.85′ W (NAD 83)

Briggs Lagoon Chart 3940;
Second Narrows, east entrance:
52°28.71′ N, 127°57.49′ W (NAD 83)

"Briggs Second Narrows" Chart 3940;
west entrance: 52°28.67′ N, 127°57.60′ W
(NAD 83)

Broad Bay Charts 3934; 3931;
anchor: 51°17.28′ N, 127°35.74′ W (NAD 83)

Broad Point West (Hyde Creek)
Sector Light Chart 3546;
position: (S. of Haddington Isl.):
50°34.97′ N, 127°01.41′ W (NAD 83)

Brockton Island Light Chart 3679;
position: (NW end of isl.):
50°29.35′ N, 127°46.29′ W (NAD 83)

Brockton Point Sector Light
Chart 3493;
position: (extremity of pt, inside First
Narrows): 49°18.07′ N, 123°06.95′ W
(NAD 83)

Broken Group Chart 3670
Meares Bluff, Effingham Island
position: 48°52.02' N, 125°17.30' W (NAD 27)

Broken Islands Chart 3545, 3564;
Broken Islands light:
50°30.67' N, 126°17.92' W;
south entrance: 50°30.70' N, 126°17.38' W
(NAD 83)

Broken Islands Light Chart 3564;
position: 50°30.67' N, 126°17.92' W (NAD 83)

Brooke Shoal Light Chart 3920;
position: (on shoal):
55°22.93' N, 129°42.90' W (NAD 83)

Brooks Bay Chart 3680
position: 50°15.00' N, 127°55.00' W (NAD 27)

Brooks Cove, Halfmoon Bay
Chart 3535 (inset);
anchor: 49°30.50' N, 123°56.68' W (NAD 27)

Brooks Peninsula Chart 3680;
See Solandar Island

Brotchie Ledge Light Chart 3415;
position: (off entrance to hbr):
48°24.40' N, 123°23.20' W (NAD 27)

Brown Bay Chart 3539 (inset);
north tip of breakwater:
50°09.80' N, 125°22.33' W (NAD 83)

Brown Bay Light Chart 3539;
position: 50°10.00' N, 125°22.03' W (NAD 83)

**Brown Channel Light
and Whistle Buoy MC** Chart 3683;
position: 49°59.49' N, 127°26.81' W (NAD 27)

Brown Cove Chart 3734;
entrance: 52°40.95' N, 128°34.40' W (NAD 27)

Brown Passage Chart 3957;
west entrance: 54° 15.19' N, 130°55.40' W
(NAD 83)

**Brown Passage Light and Whistle
Buoy D60** Chart 3957;
position: 54°18.40' N, 130°54.85' W (NAD 83)

Browning Entrance Chart 3747;
west entrance(1.0 mile north of White Rock
light): 53°39.10' N, 130°34.00' W (NAD 27)

Browning Entrance Light Chart 3747;
position: (N. isl., White Rocks):
53°38.08' N, 130°33.80' W (NAD 27)

Browning Inlet Chart 3686
entrance: 50°29.20' N, 128°03.40' W (NAD 83)

Browning Passage Chart 3549;
south entrance: 50°49.75' N, 127°38.30' W
(NAD 83)

Browning Passage Chart 3685
west entrance: 49°09.00' N, 125°53.00' W
(NAD 27)

Brownsville Light Buoy S40
Chart 3490;
position: (E. of Pattulo bridge):
49°12.72' N, 122°53.17' W (NAD 83)

Brundige Inlet Chart 3909 (inset);
entrance: 54°36.86' N, 130°50.73' W;
anchor (bitter end):
54°35.30' N, 130°53.50' W (NAD 83)

Brunswick Point Light Chart 3526;
position: (on pt, E. side of Montagu Channel):
49°31.57' N, 123°15.58' W (NAD 83)

Brydon Anchorage Chart 3784;
entrance: 51°50.25' N, 128°12.15' W;
anchor: 51°50.54' N, 128°12.05' W
(NAD 27)

Brydon Channel Chart 3784;
east entrance: 51°50.08′ N, 128°10.61′ W;
west entrance: 51°49.83′ N, 128°10.60′ W
(NAD 27)

"Brynelldson Bay" Charts 3729, 3730;
entrance: 52°26.60′ N, 127°13.00′ W;
anchor: 52°26.83′ N, 127°13.27′ W
(NAD 27)

Buccaneer Bay Marine Park
Chart 3535
(Welcome Passage inset); anchor:
49°29.71′ N, 123°59.20′ W (NAD 27)

Buchan Inlet Chart 3721;
entrance: 53°22.06′ N, 129°47.00′ W
(NAD 27)

Buchholz Rock Light Buoy M14
Chart 3679;
position: 50°29.29′ N, 127°34.40′ W
(NAD 83)

Buckley Bay Ferry Landing Light
Chart 3527;
position: (S. side of landing):
49°31.62′ N, 124°50.78′ W (NAD 27)

Buie Creek Chart 3737;
anchor (northwest corner):
52°58.20′ N, 128°40.00′ W (NAD 27)

Bull Cove Charts 3934, 3931;
position: 51°16.67′ N, 127°36.30′ W
(NAD 83)

Bull Harbor Entrance Light Chart 3549;
position: (NW corner of float):
50°54.76′ N, 127°55.72′ W (NAD 83)

Bull Harbour Chart 3549 (inset);
anchor: 50°55.05′ N, 127°56.13′ W (NAD 83)

"Bull Passage Bight" Charts 3312, 3512;
anchor: 49°28.83′ N, 124°12.45′ W (NAD 27)

Bulley Bay Chart 3734;
position: 52°28.20′ N, 128°19.10′ W (NAD 27)

Bullock Channel Chart 3940;
south entrance: 52°19.20′ N, 128°02.30′ W;
north entrance: 52°26.40′ N, 128°05.50′ W
(NAD 83)

"Bullock Channel North Cove"
Chart 3940;
anchor: 52°26.14′ N, 128°04.89′ W (NAD 83)

"Bullock Spit Cove" Chart: 3940;
anchor: 52°20.00′ N, 128°02.27′ W (NAD 83)

Bully Island Light Chart 3773;
position: (westerly side of isl.):
53°47.90′ N, 130°19.45′ W (NAD 27)

Bunsby Islands Chart 3683;
position: 50°06.00′ N, 127°32.00′ W (NAD 27)

Burdwood Bay Charts 3539, 3538;
anchor: 50°09.87′ N, 125°05.76′ W (NAD 83)

"Burdwood Point Cove" Chart 3664;
anchor: 49°34.80′ N, 126°33.50′ W (NAD 27)

Burgoyne Bay Chart 3313, p. 14;
anchor: 48°47.57′ N, 123°31.29′ W;
anchor: 48°47.38′ N, 123°31.32′ W (NAD 83)

Burial Cove Charts 3545, 3564;
anchor: 50°33.68′ N, 126°13.24′ W (NAD 83)

Burial Islet Light Chart 3470;
position: (NE corner of islet):
48°46.18′ N, 123°33.73′ W (NAD unknown)

Burke Channel Charts 3729, 3730
south entrance: 51°54.70′ N, 127°53.50′ W
(NAD 27)

Burly Bay Chart 3547;
position: 50°54.68′ N, 126°47.29′ W
(NAD 83)

Burnaby Shoal Light Chart 3493;
position: (on shoal):
49°17.92' N, 123°06.57' W (NAD 83)

Burnaby Strait Chart 3809
south entrance: 52°19.20' N, 131°19.65' W
north entrance: 52°24.20' N, 131°22.50' W
(NAD 27)

Burnett Bay Chart 3550;
position (north end):
51°08.30' N, 127°42.27' W (NAD 83)

Burns Bay , Mink Trap Bay Chart 3721;
entrance: 53°26.50' N, 129°51.60' W (NAD 27)

Burns Point Light Chart 3495;
position: (on pt, N. side of entrance to Port
Moody): 49°17.60' N, 122°55.06' W (NAD 83)

**Burrard Inlet Cautionary Light
Buoy QB** Chart 3493;
position: 49°19.04' N, 123°12.00' W (NAD 83)

Burrard Inlet Chart 3481;
cautionary light, buoy "QB":
49°19.03' N, 123°12.00' W (NAD 27)

Bush Rock Light Chart 3809;
position: (N. end of rock):
52°18.25' N, 131°16.57' W (NAD 27)

Bute Inlet Charts 3542, 3541, 3312;
entrance: 50°21.20' N, 125°06.50' W
(NAD 27)

Butedale Chart 3739 (inset);
entrance: 53°09.60' N, 128°41.50' W;
light, west end Work Island:
53°10.72' N, 128°41.55' W (NAD 27)

Butler Cove, Stephens Island
Chart 3956;
entrance: 54°06.30' N, 130°41.30' W;
mooring buoys: 54°06.73' N, 130°39.87' W
(NAD 83)

Butterworth Rocks Light Chart 3802;
position: (on rock):
54°14.13' N, 130°58.50' W (NAD 27)

Butze Rapids Chart 3958 (inset);
position: 54°18.26' N, 130°14.80' W (NAD 27)

Cabbage Island Marine Park
Chart 3313, p. 24;
anchor: 48°47.80' N, 123°05.43' W (NAD 83)

Cachalot Inlet Chart 3682
entrance: 50°00.10' N, 127°09.30' W (NAD 27)

Cadboro Bay (home of Royal Victoria
Yacht Club) Chart 3313, p. 4;
anchor: 48°27.35' N, 123°17.66' W (NAD 83)

Calamity Point Light Chart 3493;
position: 49°18.76' N, 123°07.58' W (NAD 83)

Calamity Shoal Light Buoy Q65
Chart 3493;
position: 49°18.68' N, 123°07.53' W (NAD 83)

Calm Channel Chart 3541;
anchor: 50°16.38' N, 125°02.98' W (NAD 27)

Cameron Cove Chart 3723 (inset);
anchor: 53°03.80' N, 129°06.98' W (NAD 27)

Camp Point Light Chart 3544;
position: (on pt): 50°23.08' N, 125°49.63' W
(NAD 83)

Campania Sound Charts 3737, 3724;
position: 52°58.00' N, 129°15.00' W (NAD 27)

Campbell Bay Charts 3313, p. 11;
anchor: 48°51.53' N, 123°16.20' W (NAD 83)

Campbell Cove Charts 3641, 3440, 3430;
anchor: 48°19.67' N, 123°37.83' W (NAD 27)

"Campbell Island Inlet" Chart 3787;
entrance: 52°02.93' N, 128°10.92' W;
anchor (Wendy Cove):
52°05.36' N, 128°12.59' W (NAD 27)

Campbell River Breakwater Light
Chart 3540;
position: (on breakwater extension):
50°01.48' N, 125°14.26' W (NAD 83)

Campbell River Chart 3540;
Discovery Harbour Marina breakwater light:
50°02.16' N, 125°14.49' W;
Campbell River breakwater light:
50°01.48' N, 125°14.26' W (NAD 83)

**Campbell River Ferry Terminal
Dolphin No 2 Light** Chart 3540;
position: 50°01.83' N, 125°14.27' W (NAD 83)

**Campbell River Ferry Terminal
Dolphin No 3 Light** Chart 3540;
position: 50°01.75' N, 125°14.28' W (NAD 83)

Campbell River North Light
Chart 3540;
position: 50°01.71' N, 125°14.38' W (NAD 83)

Canal Bight Chart 3785;
entrance: 52°05.30' N, 128°05.10' W;
anchor: 52°05.48' N, 128°05.20' W (NAD 27)

Canal Island Light Chart 3664;
position: (on W. side of isl.):
49°41.30' N, 126°35.06' W (NAD 27)

Cannery Bay Chart 3649;
anchor: 49°08.36' N, 125°40.27' W (NAD 27)

**Cannery Channel,
Steveston Harbour** Chart 3490;
Cannery Channel anchor:
49°06.88' N, 123°09.46' W (NAD 83)

Canoe Bay ("Canoe Cove")
Chart 3313, p. 7;
outer float: 48°40.95' N, 123°24.03' W
(NAD 83)

Canoe Cove Chart 3934:
position: 51°28.02' N, 127°52.68' W (NAD 83)

Canoe Island Chart 3670;
position: 48°57' N, 125°15' W (NAD 83)

Canoe Pass Light and Bell Buoy T14
Chart 3463;
position: (entrance to pass):
49°02.30' N, 123°15.30' W (NAD 27)

Canoe Passage Chart 3463;
Canoe Passage light and bell buoy "T14":
49°02.30' N, 123°15.30' W (NAD 27)

Canoe Rock Light Chart 3476;
position: (on the rock):
48°44.02' N, 123°20.33' W (NAD 27)

Cape Beale Sector Light Chart 3671;
position: (SE pt of Barkley Sound entrance):
48°47.20' N, 125°12.85' W (NAD 27)

Cape Calvert Chart 3934;
Clark Point light: 51°25.80' N, 127°53.20' W
(NAD 83)

Cape Caution Chart 3551;
position (10 fathom patch 1.3 miles SE of
point): 51°09.27' N, 127°48.93' W (NAD 83)

Cape Caution Light Chart 3550;
position: (on the cape):
51°09.83' N, 127°47.10' W (NAD 83)

Cape Cockburn Light Chart 3311;
position: (on cape W side of Nelson Isl.):
49°40.26' N, 124°12.09' W (NAD 83)

Cape Cook Chart 3680;
position: 50°08' N, 127°55' W (NAD 27)

"Cape Cook Lagoon" Chart 3680;
entrance: 50°11.95' N, 127°48.13' W
(NAD 27)

Cape Farewell Light Chart 3711;
position: 53°21.46' N, 129°13.64' W
(NAD 27)

Cape Lazo Chart 3527;
Cape Lazo light: 49°42.41′ N, 124°51.69′ W
(NAD 27)

Cape Lazo Light Chart 3527;
position: (on northern part of cape):
49°42.41′ N, 124°51.69′ W (NAD 27)

Cape Mark Light Chart 3787;
position: 52°08.98′ N, 128°32.30′ W (NAD 27)

Cape Mudge Charts 3539, 3540;
light: 49°59.93′ N, 125°11.63′ W (NAD 83)

Cape Mudge Sector Light Chart 3540;
position: (W. extremity):
49°59.91′ N, 125°11.74′ W (NAD 83)

Cape Roger Curtis Light Chart 3526;
position: (on the pt):
49°20.40′ N, 123°25.88′ W (NAD 83)

Cape Scott Light Chart 3598;
position: (NW tip Vanc. Isl.):
50°46.95′ N, 128°25.55′ W (NAD 27)

Cape Scott to Lippy Point Chart 3624

Cape St. James Light Chart 3825;
position: (on cape, St. James Isl.):
51°56.17′ N, 131°00.87′ W (NAD 27)

Cape Sutil Chart 3549;
anchor: 50°52.20′ N, 128°02.90′ W (NAD 83)

Captain Cove Charts 3753 (inset);
entrance: 53°48.73′ N, 130°13.05′ W;
anchor: 53°48.65′ N, 130°11.80′ W (NAD 27)

Captain Island Light Buoy Chart 3514;
position: (W. of island):
49°47.70′ N, 124°02.10′ W (NAD 83)

Captain Island Light Chart 3514;
position: (northwesterly shore of isl.):
49°47.36′ N, 123°59.65′ W (NAD 83)

Captain Passage Light Buoy U62
Chart 3478;
position: (SE of Annette Pt):
48°49.72′ N, 123°24.30′ W (NAD 83)

Car Point Notch Chart 3933;
position: 55°20.30′ N, 130°00.00′ W (NAD 27)

Carlson Inlet Chart 3730;
entrance: 52°33.80′ N, 127°13.00′ W;
head of inlet: 52°34.80′ N, 127°24.30′ W
(NAD 27)

Carmanah Light Chart 3606;
position: (on pt): 48°36.72′ N, 124°45.00′ W
(NAD 27)

Carmanah Point Chart 3606;
light: 48°36.72′ N, 124°44.00′ W

Carmichael Passage Chart 3894;
north entrance: 53°00.85′ N, 131°56.50′ W;
south entrance: 53°56.10′ N, 131°54.10′ W
(NAD 27)

Carolina Channel Charts 3646, 3671;
entrance buoys (from Light List)—
bouy "Y42": 48°54.72′ N, 125°32.54′ W;
buoy "Y43": 48°55.18′ N, 125°31.67′ W;
buoy "Y46": 48°55.68′ N, 125°31.20′ W

**Carolina Channel Inner Light
and Bell Buoy Y43** Chart 3646;
position: 48°55.18′ N, 125°31.67′ W
(NAD 83)

**Carolina Channel Light
and Whistle Buoy Y42** Chart 3646;
position: (off Amphitrite Pt):
48°54.72′ N, 125°32.54′ W (NAD 83)

"Carolina Island Anchorage"
Chart 3955;
entrance: 54°20.24′ N, 130°25.52′ W;
anchor: 54°20.47′ N, 130°25.72′ W (NAD 27)

Carraholly Point Light Chart 3495;
position: (on rock, off pt):
49°17.62' N, 122°54.51' W (NAD 83)

Carriden Bay Chart 3547;
anchor: 50°54.42' N, 126°54.53' W (NAD 83)

Carrie Bay Chart 3546;
anchor: 50°41.64' N, 126°37.86' W (NAD 83)

Carrington Bay Charts 3538, 3312;
position lagoon bar:
50°08.15' N, 125°00.05' W (NAD 27)

Carter Bay Chart 3738;
entrance: 52°49.35' N, 128°23.80' W;
anchor: 52°49.82' N, 128°23.35' W (NAD 27)

Carter Passage Chart 3547;
anchor: 50°50.32' N, 126°49.02' W (NAD 83)

Carterer Point Light Chart 3544;
position: (E. end of Hardwicke Isl.):
50°27.53' N, 125°45.83' W (NAD 83)

Cartwright Bay Chart 3547;
anchor: 50°52.83' N, 126°46.50' W (NAD 83)

Cascade Harbour Chart 3549;
anchor: 50°54.42' N, 127°44.35' W (NAD 83)

Cascade Inlet Chart 3729;
entrance: 52°24.60' N, 127°24.60' W;
anchor: 52°36.40' N, 127°37.20' W (NAD 27)

Casey Cove Chart 3958;
position: 54°16.84' N, 130°22.58' W (NAD 27)

Casey Point Light Chart 3957;
position: (edge of shoal, off pt):
54°16.43' N, 130°21.68' W (NAD 83)

Catala Passage Charts 3733A, 3728;
west entrance: 52°16.25' N, 128°44.20' W;
east entrance: 52°15.65' N, 128°40.90' W
(NAD 27)

"Catala Spit Bight" Chart 3663;
anchor: 49°50.62' N, 127°02.67' W (NAD 27)

"Cataract Creek Cove" Chart 3670;
anchor: 49°01.18' N, 125°17.04' W (NAD 83)

"Cathedral Point Cove" Chart 3729;
anchor: 52°11.24' N, 127°28.00' W (NAD 27)

Cattle Islands Chart 3548;
anchor: 50°42.70' N, 127°24.45' W (NAD 83)

Caulfeild Cove Chart 3481;
anchor: 49°20.27' N, 123°15.17' W (NAD 27)

Caution Cove Chart 3545;
anchor: 50°35.15' N, 126°30.30' W (NAD 83)

Cavin Cove Charts 3785, 3787;
entrance: 52°10.60' N, 128°08.05' W (NAD 27)

Cecil Cove Chart 3807;
entrance: 52°51.30' N, 131°51.50' W;
anchor: 52°51.64' N, 131°52.15' W (NAD 27)

Cecil Island Chart 3547;
position: 50°50.90' N, 126°42.66' W (NAD 83)

Cecil Patch Light and Bell Buoy D9
Chart 3717;
position: (NE end of patch):
54°03.80' N, 130°16.58' W (NAD 83)

Celia Reefs Light Buoy U14
Chart 3476;
position: (S. of reefs):
48°42.65' N, 123°22.88' W (NAD 27)

Centre Island Light Chart 3663;
position: (NW shore of isl.):
49°50.95' N, 126°56.07' W (NAD 27)

Chained Islands Chart 3539;
anchor west: 50°14.89' N, 125°21.57' W;
anchor east: 50°14.83' N, 125°20.80' W
(NAD 83)

Chalmers Anchorage Chart 3927;
position: 54°02.75' N, 130°16.35' W (NAD 27)

Channe Passage Chart 3543;
position: 50°27.00' N, 125°19.70' W
(NAD 27)

Channel Islands Light Chart 3478;
position: (N. end of N. Channel Islands):
48°48.12' N, 123°22.85' W (NAD 83)

"Chappell Cove" Chart 3921;
anchor (north of spit):
51°04.68' N, 127°29.33' W;
anchor (head of inlet):
51°04.80' N, 127°28.53' W (NAD 83)

"Chapple Inlet Lagoon"
Charts 3719 (inset), 3737;
narrows fairway: approximately
52°57.78' N, 129°08.08' W (NAD 27)

Chapple Inlet, Doig Anchorage Charts
3719 (inset), 3737;
entrance: 52°54.00' N, 129°07.90' W;
anchor: 52°55.28' N, 129°07.80' W (NAD 27)

Charles Bay Chart 3543;
anchor: 50°25.09' N, 125°29.15' W (NAD 27)

Charlotte Bay Chart 3552;
anchor: 51°03.55' N, 127°22.40' W (NAD 27)

Chatham Channel Chart 3564;
range lights east side:
50°34.82' N, 126°12.02' W;
range light west side:
50°34.78' N, 126°14.47' W (NAD 83)

Chatham Channel Chart 3564;
north entrance: 50°37.80' N, 126°18.70' W
(NAD 83)

Chatham Point Chart 3539;
light: 50°20.02' N, 125°26.34' W (NAD 83)

Chatham Point Light Chart 3539;
position: (off pt): 50°20.01' N, 125°26.43' W
(NAD 83)

Chatham Sound Chart 3957;
position: 54°08.00' N, 130°19.00' W (NAD 27)

**Chatham Sound Light and
Bell Buoy D77** Chart 3957;
position: (S. of Lucy Isl.):
54°15.48' N, 130°38.45' W (NAD 83)

Chatterbox Falls Chart 3312;
public float: 50°12.28' N, 123°46.02' W
(NAD 27)

Cheamly Passage Light Chart 3984;
position: (SE end of passage):
54°01.57' N, 130°41.10' W (NAD unknown)

Chemainus Bay Chart 3475 (inset);
public docks: 48°55.54' N, 123°42.76' W
(NAD 27)

Chemainus Bay Range Light
Chart 3475;
position: (N. end of McMilland and Bloedel
wharf): 48°55.48' N, 123°42.74' W (NAD 27)

"Chettleburgh Point Cove"
Charts 3719 (inset), 3737;
anchor: 52°56.58' N, 129°08.57' W (NAD 27)

Chief Mathews Bay Chart 3745;
entrance: 53°22.50' N, 128°03.50' W (NAD 27)
anchor: 53°20.14' N, 128°06.00' W (NAD 27)

Chief Nollis Bay Chart 3552;
position: 51°11.05' N, 127°05.45' W (NAD 27)

Chismore Passage
Charts 3956, 3717 (partial), 3927;
north entrance: 54°04.40' N, 130°20.62' W
(NAD 83)

Choked Passage Chart 3784;
north entrance: 51°41.55' N, 128°05.80' W
(NAD 27)

Cholberg Point Light Chart 3549; position: (on pt): 50°54.95′ N, 127°44.07′ W (NAD 83)

Chonat Bay Chart 3537; position: 50°17.50′ N, 125°18.50′ W (NAD 27)

Christie Bay Chart 3891; entrance: 53°12.60′ N, 132°13.25′ W; anchor: 53°12.32′ N, 132°12.85′ W (NAD 83)

"Christie Cove" (Christie Creek) Chart 3526; 49°32.14′ N, 123°27.23′ W (NAD 27)

Chrow Islands Light Chart 3671; position: (on NW extremity of N. Isl.): 48°54.60′ N, 125°28.26′ W (NAD 27)

Chup Point Light Chart 3668; position: (E. side of pt): 48°57.32′ N, 125°01.70′ W (NAD 83)

Church House Chart 3541; public float: 50°20.13′ N, 125°04.54′ W (NAD 27)

Cigarette Cove Chart 3670; anchor: 49°00.44′ N, 125°17.42′ W (NAD 83)

Cinque Islands Light Chart 3537; position: (W. side of isl.): 50°17.73′ N, 125°23.98′ W (NAD 27)

City Reach Range 1 Light Chart 3490; position: 49°09.19′ N, 122°57.13′ W (NAD 83)

City Reach Range 2 Light Chart 3490; position: (S. bank of river): 49°09.30′ N, 122°56.68′ W (NAD 83)

Clam Bay, Kuper Island Chart 3313, p. 16; anchor: 48°58.90′ N, 123°38.55′ W (NAD 83)

Clam Inlet Chart 3959; entrance: 54°29.95′ N, 130°47.30′ W; anchor: 54°29.48′ N, 130°47.19′ W (NAD 83)

Clanninick Cove Chart 3683: anchor: 50°52.09′ N, 127°24.65′ W (NAD 27)

Clark Point Light Chart 3934; position: (on pt, Calvert Isl.): 51°25.79′ N, 127°53.20′ W (NAD 83)

Clarke Cove Charts 3737, 3724; entrance: 52°58.48′ N, 129°11.80′ W; anchor: 52°58.00′ N, 129°10.88′ W (NAD 27)

Clarke Island Chart 3670; anchor: 48°53.57′ N, 125°22.33′ W (NAD 83)

Clarke Rock Light Chart 3458; position: (E. side of rock): 49°13.53′ N, 123°56.42′ W (NAD 83)

Clarke-Benson Passage Chart 3670; light: 48°53.161′ N, 125°22.643′ W (NAD 83)

Clatse Bay Chart 3940; entrance: 52°21.80′ N, 127°51.40′ W; anchor: 52°20.53′ N, 127°50.54′ W (NAD 83)

Claydon Bay Chart 3547; anchor: 50°56.28′ N, 126°53.36′ W (NAD 83)

Clear Passage Chart 3682; "M38" buoy: 49°58.23′ N, 127°15.03′ W (NAD 27)

Clear Passage Chart 3746; anchor (east entrance): 53°33.20′ N, 129° 58.35′ W (NAD 27)

Clement Rapids Chart 3742; anchor (near Salmon Point): 53°12.63′ N, 129°03.02′ W (NAD 27)

Clerke Peninsula Light Chart 3664; position: (on S. extremity of peninsula): 49°36.16′ N, 126°32.21′ W (NAD 27)

Clerke Point Chart 3683;
way-point: 50°03.46' N, 127°48.30' W
(NAD 27)

"Clerke Reef Passage" Chart 3680;
north entrance way-point:
50°13.40' N, 127°49.96' W (NAD 27)

Cliff Point Light Chart 3686;
position: (extremity of point):
50°27.86' N, 127°56.21' W (NAD 83)

Clifford Bay Chart 3710 (inset);
entrance: 52°35.65' N, 129°09.55' W;
anchor (west side of Craft Island):
52°35.36' N, 129°08.98' W (NAD 27)

Clifton Point Chart 3671;
anchor: 48°55.30' N, 125°03.50' W (NAD 27)

Clio Bay Chart 3743;
entrance: 53°54.75' N, 128°41.70' W;
anchor: 53°53.74' N, 128°40.07' W (NAD 27)

Clio Channel Chart 3545;
north entrance: 50°37.75' N, 126°20.40' W
(NAD 83)

Clio Channel Chart 3546;
north entrance: 50°37.75' N, 126°20.40' W
(NAD 83)

Clio Island Light Chart 3648;
position: (on W. end of isl.):
49°23.63' N, 126°10.95' W (NAD 27)

Clio Point Light Chart 3743;
position: (on pt): 53°54.40' N, 128°42.50' W
(NAD 27)

Clive Island Light Chart 3476; position;
48°42.08' N, 123°24.23' W (NAD 27)

Clo-oose Light and Whistle Buoy YJ
Chart 3606;
position: (off Clo-oose):
48°38.83' N, 124°49.83' W (NAD 27)

Coal Harbour Chart 3679;
wharf: 50°35.88' N, 127°34.80' W (NAD 83)

Coal Harbour Light Chart 3493;
position: 49°17.62' N, 123°07.47' W (NAD 83)

Coal Harbour, Port of Vancouver
Chart 3493;
entrance: 49°17.60' N, 123°07.00' W (NAD 83)

Coal Island Light Buoy U8 Chart 3476;
position: (N. of island):
48°41.63' N, 123°23.03' W (NAD 27)

Coal Island Light Chart 3476;
position: (on Fir Cone Pt):
48°41.48' N, 123°23.21' W (NAD 27)

Coaster Channel Light Chart 3670;
position: 48°53.00' N, 125°19.07' W (NAD 83)

Cockburn Bay Chart 3311;
entrance: 49°40.85' N, 124°12.05' W (NAD 83)

Cockle Bay Chart 3728;
anchor: 52°20.75' N, 128°23.25' W (NAD 27)

Codville Lagoon Chart 3785;
entrance: 52°03.30' N, 127°53.00' W;
anchor: 52°03.68' N, 127°50.15' W (NAD 27)

Coffin Island Light Chart 3475;
position: (on islet, N. side of entrance to
Ladysmith Hbr): 48°59.15' N, 123°45.14' W
(NAD 27)

Coghlan Anchorage
Charts 3711 (inset), 3742;
south entrance: 53°21.90' N, 129°15.80' W;
north entrance: 53°24.70' N, 129°14.75' W;
Harbour Rock Light:
53°23.29' N. 129°16.54' W;
anchor (Otter Shoal):
53°23.82' N, 129°17.12' W (NAD 27)

Coho Cove Chart 3313, p. 20 (inset);
position: 49°08.25' N, 123°48.45' W
(NAD 83)

Colburne Passage Light Buoy U18
Chart 3476;
position: (S. of Piers Isl.):
48°41.90' N, 123°25.23' W (NAD 27)

Colburne Passage South Light
Chart 3476;
position: 48°41.85' N, 123°25.47' W
(NAD 27)

Colby Bay Chart 3753 (inset);
entrance: 53°32.15' N, 130°10.00' W;
anchor: 53°31.82' N, 130°10.52' W (NAD 27)

Coles Bay Chart 3313, p. 13;
position: 48°38.90' N, 123°32.95' W (NAD 83)

Colliery Range Light Chart 3457;
position: 49°09.95' N, 123°55.72' W (NAD 27)

Collins Bay Chart 3743;
position: 53°33.20' N, 128°44.20' W (NAD 27)

**Colquhoun Shoal North Cardinal
Light and Whistle Buoy DCO**
Chart 3957;
position: 54°09.45' N, 130°30.85' W (NAD 83)

Columbia Cove ("Peddlers Cove")
Chart 3683;
anchor: 50°08.33' N, 127°41.48' W (NAD 27)

Commando Inlet Chart 3719 (inset);
entrance: 52°47.14' N, 129°06.14' W;
anchor (reported at east end):
52°48.07' N, 129°04.37' W (NAD 27)

Commerell Point Chart 3624;
anchor: 50°34.47' N, 128°14.47' W (NAD 27)

Comox Aeronautical Light Chart 3527;
position: 49°42.35' N, 124°53.07' W
(NAD 27)

Comox Bar Charts 3527 or 3513;
bar position: about 49° 39'N, 124°53'W
(NAD 27)

Comox Bar Light and Bell Buoy P54
Chart 3527;
position: (close to easterly edge of bar):
49°39.50' N, 124°51.60' W (NAD 27)

Comox Bar Range Light Chart 3527;
position: (on W. shore of Baynes Sound):
49°37.40' N, 124°54.45' W (NAD 27)

Comox Harbour Breakwater Light
Chart 3527;
position: 49°40.17' N, 124°55.75' W
(NAD 27)

Comox Harbour Chart 3527 (inset);
breakwater light: 49°40.17' N, 124°55.71' W
(NAD 27)

**Comox Harbour East
Breakwater Light** Chart 3527;
position: (outer end of new breakwater):
49°40.08' N, 124°55.41' W (NAD 27)

"Concepcion Point Bay" Chart 3664;
entrance: 49°39.45' N, 126°29.30' W
(NAD 27)

Conconi Reef Light Chart 3442;
position: (on reef):
48°49.45' N, 123°17.46' W (NAD 27)

Connis Cove Charts 3747, 3927;
entrance: 53°45.40' N, 130°18.00' W;
anchor: 53°45.20' N, 130°17.77' W (NAD 27)

Connis Islet Light Chart 3747;
position: (on islet, Beaver Passage):
53°45.46' N, 130°18.98' W (NAD 27)

Conover Cove Chart 3313, p.17;
anchor: 48°56.13' N, 123°32.55' W
(NAD 83)

Conville Bay Chart 3539;
position: 50°10.80' N, 125°09.00' W
(NAD 83)

Convoy, Patrol, Fairmile, Souvenir and Sweeper Island Passages Chart 3921;
south entrance (Convoy Passage):
51°34.80′ N, 127°48.90′ W;
west entrance (Patrol Passage):
51°36.90′ N, 127°51.20′ W;
west entrance (Fairmile Passage):
51°37.80′ N, 127°50.20′ W;
east entrance (Souvenir Passage):
51°37.35′ N, 127°48.60′ W (NAD 83)

Cook Point Light Chart 3893;
position: (Masset Inlet entrance):
53°48.38′ N, 132°12.10′ W (NAD 27)

Coolidge Point Light Chart 3781;
position: (on pt, Cousins Inlet):
52°21.03′ N, 127°42.92′ W (NAD 27)

Coomes Bank Light Buoy Y11
Chart 3648;
position: (off S. end of bank):
49°13.37′ N, 126°00.42′ W (NAD 27)

Cooper Inlet Chart 3785;
entrance: 52°04.00′ N, 128°04.25′ W (NAD 27)

Cooper Reach: McBride Bay, Frazer Bay Chart 3543 (inset);
McBride Bay position:
50°42.95′ N, 125°26.05′ W;
Frazer Bay position:
50°43.00′ N, 125°27.70′ W′ (NAD 27)

Copeland Islands Marine Park
Charts 3311, 3538;
anchor south end:
50°00.06′ N, 124°48.13′ W (NAD 83)

Cordero Channel Chart 3543
west entrance: 50°26.45′ N, 125°35.90′ W
east entrance: 50°27.10′ N, 125°18.80′ W
(NAD 27)

"Cordero Islands Cove"
Chart 3543 (inset);
anchor: 50°26.73′ N, 125°29.78′ W (NAD 27)

Cordova Bay Chart 3313, p. 5;
anchor: 48°29.77′ N, 123°19.12′ W (NAD 83)

Corney Cove Chart 3710 (inset);
entrance: 52°36.50′ N, 128°42.60′ W;
anchor: 52°36.93′ N, 128°42.68′ W (NAD 27)

Cornwall Inlet and Drake Inlet
Chart 3742;
entrance: 53°12.83′ N, 129°03.21′ W (NAD 27)

Cornwall Point Light Chart 3745;
position: (on pt): 53°28.65′ N, 128°21.62′ W
(NAD 27)

Cortes Bay Chart 3538, 3311 (inset);
public float: 50°03.73′ N, 124°55.06′ W
(NAD 83)

Cortes Bay Light Chart 3311;
position: 50°03.79′ N, 124°55.42′ W (NAD 83)

Cougar Bay Charts 3738, 3734;
entrance: 52°43.75′ N, 128°34.70′ W;
anchor: 52°44.52′ N, 128°34.60′ W (NAD 27)

Cougar Inlet Chart 3921;
entrance: 51°05.18′ N, 127°30.56′ W (NAD 83)

Coulter Bay, Cortes Island
Charts 3538; 3312;
anchor: 50°07.95′ N, 125°02.65′ W (NAD 27)

Courtenay River and Slough
Chart 3527 (inset)
position: 49°40.97′ N, 124°58.94′ W
(NAD 27)

Courtenay River Approach Light
Chart 3527;
position: (river entrance):
49°39.95′ N, 124°56.63′ W (NAD 27)

Courtenay River Range Light
Chart 3527;
position: (on bank of river):
49°40.64′ N, 124°57.46′ W (NAD 27)

Cousins Inlet Charts 3781, 3720, 3729;
entrance position:
52°16.10' N, 127°46.20' W (NAD 27)

Cousins Inlet Light Chart 3781;
position: 52°19.65' N, 127°44.90' W (NAD 27)

"Cove East of Cheenis Lake"
Chart 3940;
anchor: 52°30.88' N, 128°06.07' W (NAD 83)

"Cove East of Jansen Lake"
Chart 3682,
anchor: 50°06.2' N, 127°16.4' W (NAD 27)

"Cove East of Westcott Point"
Chart 3548;
entrance: 50°58.25' N, 127°27.30' W;
anchor: 50°58.52' N, 127°27.27' W (NAD 83)

"Cove Northeast of Columbia Cove"
Chart 3683;
anchor: 50°08.93' N, 127°40.38' W (NAD 27)

"Cove Northwest of Arthur Island"
Chart 3734;
east entrance: 52°27.32' N, 128°15.80' W;
anchor: 52°27.49' N, 128°16.37' W (NAD 27)

"Cove Southeast of Latta Island"
Chart 3786;
entrance: 51°58.60' N, 128°11.75' W (NAD 27)

Cow Bay (PRRYC) Chart 3958;
breakwater position:
54°19.73' N, 130°19.12' W (NAD 83)

Cowards Cove Chart 3481;
anchor: 49°15.37' N, 123°15.82' W (NAD 27)

"Cowards Cove" Charts 3737, 3726;
entrance: 52°33.60' N, 128°44.65' W;
anchor (east of lagoon):
52°34.00' N, 128°44.34' W (NAD 27)

Cox Island Chart 3624;
position: 50°48' N, 128°36' W (NAD 27)

Crab Cove Chart 3784;
anchor (north): 51°50.87' N, 128°01.28' W
(NAD 27)

"Crab Trap Cove" Charts 3761, 3927;
south entrance: 53°50.97' N, 130°30.91' W;
north entrance: 53°51.80' N, 130°31.40' W;
anchor: 53°51.07' N, 130°30.00' W (NAD 27)

Cracroft Inlet Chart 3545;
north entrance: 50°36.08' N, 126°21.02' W
(NAD 83)

**Cracroft Point and Blackney Passage
to Blackfish Sound** Chart 3546;
Blackney Passage, Licka Point light:
50°33.98' N, 126°41.48' W (NAD 83)

Cracroft Point Light Chart 3546;
position: (Blackney Passage entrance):
50°32.99' N, 126°40.75' W (NAD 83)

Craig Bay Chart 3459;
position: 49°18.77' N, 124°14.55' W (NAD 27)

Crane Bay Chart 3742;
entrance: 53°13.75' N, 129°18.50' W (NAD 27)

Crane Islands Light Chart 3549;
position: (on summit westerly islet):
50°50.53' N, 127°31.33' W (NAD 83)

"Craven Hill Bight" Chart 3544;
position: 50°27.12' N, 125°54.68' W (NAD 83)

Crawford Anchorage Chart 3543;
anchor: 50°26.12' N, 125°27.66' W (NAD 83)

"Crease Island Cove" Chart 3546;
anchor: 50°36.94' N, 126°38.60' W (NAD 83)

Crescent Beach Chart 3463;
position: 49°03.42' N, 122°52.05' W (NAD 27)

Crescent Beach Light Chart 3463;
position: (entrance to Nicomeki R.):
49°00.75' N, 122°56.18' W (NAD 27)

Crescent Channel & Bold Point
Chart 3539;
anchor: 50°10.30' N, 125°10.05' W (NAD 83)

Crescent Channel Light Chart 3463;
position: 49°03.25' N, 122°53.75' W
(NAD 27)

Crescent Inlet Chart 3807;
entrance: 52°45.00' N, 131°47.90' W;
anchor: 52°44.85' N, 131°52.83' W (NAD 27)

Crescent Island Light Buoy S60
Chart Nil;
position: (upstream from isl.):
49°10.00' N, 122°26.00' W (NAD unknown)

Cricket Bay Chart 3671;
position: 48°65.41' N, 125°01.13' W
(NAD 27)

**Crispin Rock Bifurcation Light
Buoy UJ** Chart 3477;
position: (Lyall Hbr):
48°48.07' N, 123°11.78' W (NAD 27)

"Critter Cove" Chart 3664;
anchor: 49°42.85' N, 126°30.62' W (NAD 27)

Crofton Light Chart 3475;
position: (on SE Shoal Isl.):
48°52.78' N, 123°37.72' W (NAD 27)

Croker Island South Light Chart 3495;
position: 49°25.84' N, 122°51.83' W
(NAD 83)

Croker Point Light Chart 3477;
position: (on the pt):
48°46.48' N, 123°12.08' W (NAD 27)

Crow Lagoon Chart 3994; outer
entrance: 54°42.90' N, 130°12.80' W (NAD 27)

Croyden Bay Charts 3730, 3729;
anchor: 52°18.68' N, 127°09.56' W (NAD 27)

Cullen Harbour Chart 3547;
entrance: 50°45.94' N, 126°44.55' W
(NAD 83)

Culpepper Lagoon Chart 3962 (inset);
entrance: 52°45.16' N, 127°52.97' W;
anchor (lagoon head):
52°43.95' N, 127°49.75' W (NAD 27)

"Cultus Bay" Chart 3786;
anchor: 51°54.44' N, 128°11.56' W (NAD 27)

Cultus Sound Chart 3786;
southwest entrance:
51°54.08' N, 128°14.31' W (NAD 27)

Cumshewa Inlet and Approach
Chart 3894;
entrance: 53°01.40' N, 131°35.50' W (NAD 27)

Cunningham Passage
Charts 3959, 3963;
south entrance (0.10 mile north of buoy
"DK"): 54°30.90' N, 130°28.40' W;
north entrance: 54°34.40' N, 130°28.40' W
(NAD 83)

Curlew Bay Chart 3742;
entrance: 53°17.10' N, 129°18.53' W;
anchor (west side):
53°16.71' N, 129°19.32' W (NAD 27)

Current Passage Sector Light
Chart 3544;
position: (S. side of Hardwicke Isl.):
50°24.55' N, 125°48.62' W (NAD 83)

Currie Islet (Gosling Rocks) Light
Chart 3786;
position: (on islet):
51°51.10' N, 128°27.30' W (NAD 27)

Curtis Inlet Chart 3746 (inset);
entrance: 53°30.09' N, 129°53.66' W (NAD 27)

Cutter Cove Charts 3564, 3545;
anchor: 50°37.28' N, 126°15.35' W NAD 83)

"Cypress Cove" Chart 3649;
anchor: 49°16.83' N, 125°52.93' W (NAD 27)

Cypress Harbour, Berry Cove
Chart 3547;
anchor: 50°50.13' N, 126°40.30' W (NAD 83)

Cypress Harbour, Miller Bay
Chart 3547;
anchor: 50°50.24' N, 126°39.55' W (NAD 83)

Cyril Rock Light Chart 3311;
position: (on the rock, off Grisle Pt):
49°48.35' N, 124°36.33' W (NAD 83)

D'Arcy Island Light Chart 3441;
position: (SW side of isl.):
48°33.94' N, 123°17.02' W (NAD 27)

D'Arcy Island Marine Park
Chart 3313, p. 5;
anchor: 48°33.80' N, 123°17.00' W (NAD 83)

D'Arcy Shoals Light Buoy U1
Chart 3441;
position: 48°34.23' N, 123°18.03' W
(NAD 27)

Dall Rocks Light and Bell Buoy E33
Chart 3787;
position: (off N. extremity of rocks):
52°13.05' N, 128°09.28' W (NAD 27)

Dallwood ODAS Light Buoy 46207
Chart 3744;
position: 50°51.60' N, 129°54.60' W
(NAD 27)

Dana Passage Chart 3807; north
entrance: 52°50.25' N, 131°51.00' W;
south entrance: 52°49.10' N, 131°49.80' W;
anchor (center of passage):
52°49.80' N, 131°50.35' W (NAD 27)

Danger Reefs Light Chart 3443;
position: (N. end of easternmost rock):
49°03.26' N, 123°42.83' W (NAD 27)

Darby Channel Chart 3934;
west entrance: 51°30.80' N, 127°45.40' W;
east entrance: 51°34.50' N, 127°35.00' W
(NAD 83)

Dark Cove Chart 3312;
position: 49°48.03' N, 123°57.75' W
(NAD 27)

Darwin Point Light Chart 3808;
position: (NW of pt):
52°34.53' N, 131°27.70' W (NAD 27)

Davey Rock Light Buoy N32
Chart 3549;
position: 50°51.62' N, 127°31.07' W
(NAD 83)

David Bay Chart 3962;
entrance: 52°53.60' N, 128°07.30' W;
anchor: 52°52.50' N, 128°07.20' W (NAD 27)

David Cove Chart 3313, p. 11;
position: 48°51.96' N, 123°16.42' W
(NAD 83)

Davidson Point Light Chart 3859;
position: (N. side of entrance to Tasu Sound):
52°44.53' N, 132°06.70' W (NAD 27)

Davie Bay Charts 3512, 3513;
position: 49°36.18' N, 124°23.62' W (NAD 27)

Davis Bay Chart 3547;
anchor: 50°53.33' N, 127°00.83' W (NAD 83)

Dawley Passage Chart 3685;
entrance: 49°09' N, 125°48' W (NAD 83)

Dawson Harbour Chart 3891 (inset);
entrance: 53°09.75' N, 132°29.10' W;
Dawson Harbour mooring buoy:
53°10.27' N, 132°26.48' W (NAD 83)

Dawson Islands Light Chart 3893;
position: (on eastermost isl.):
53°42.98' N, 132°20.17' W (NAD 27)

Dawsons Landing Chart 3934;
float: 51°34.48' N, 127°35.50' W (NAD 83)

De Cosmos Lagoon Chart: 3785;
position: 52°56.30' N, 127 57.88' W
(NAD 27).

De Freitas Islets Chart 3734;
entrance: 52°27.15' N, 128°14.00' W;
anchor: 52°26.85' N, 128°14.90' W (NAD 27)

De Horsey Island Light Chart 3717;
position: (E. side of isl.):
54°07.87' N, 130°07.32' W (NAD 83)

"De la Beche Cove" Chart 3808; east
entrance: 52°32.15' N, 131°37.85' W;
anchor: 52°32.18' N, 131°38.36' W (NAD 27)

De la Beche Inlet *Chart* 3808;
entrance: 52°32.35' N, 131°38.70' W
(NAD 27)

"Dead Point Cove" Chart 3545;
anchor: 50°35.53' N, 126°35.17' W (NAD 83)

**Dead Tree Point Light and Whistle
Buoy C19** Chart 3890;
position: (E. of pt): 53°20.58' N,
131°53.60' W (NAD 27)

Deadman Island Light Chart 3493;
position: (south of isl.):
49°17.57' N, 123°07.33' W (NAD 83)

Deadman Islets Light Chart 3685;
position: (on the southerly islet):
49°09.74' N, 125°54.42' W (NAD 83)

Dean Channel Chart 3781;
west entrance: 52°15.00' N, 127°46.00' W
(NAD 27)

Dean Point Light Chart 3541;
position: (West Redonda Isl.):
50°17.15' N, 124°47.12' W (NAD 83)

"Dearth Island Cove" Chart 3720;
anchor (south): 52°14.98' N, 127°56.58' W
(NAD 27)

Deas Island Range Light Chart 3490;
position: (W. end of isl.):
49°07.12' N, 123°04.47' W (NAD 83)

Deas Point Light Chart 3729;
position: (flat rock adjoining headland,
Labouchere Channel):
52°23.13' N, 127°12.93' W (NAD 27)

Deas Slough Light Chart 3490;
position: (entrance to dredged cut):
49°06.92' N, 123°04.85' W (NAD 83)

Deep Bay Charts 3527, 3513;
light on southwest end of Mapleguard Point:
49°27.98' N, 124°44.07' W (NAD 27)

Deep Bay Light Chart 3527;
position: (SW end of Mapleguard Pt):
49°27.98' N, 124°44.07' W (NAD 27)

"Deep Bay" Charts 3312, 3512,
anchor: 49 30.35'N, 124 12.80'W (NAD 27)

Deep Cove Chart 3495;
position: 49°19.70' N, 122°56.70' W (NAD 83)

Deep Harbour Chart 3515;
entrance: 50°47.40' N, 126°37.00' W
(NAD 83)

**Deep Ridge Bifurcation Light
Buoy UN** Chart 3478;
position: (SE of Channel Isl.):
48°47.83' N, 123°22.10' W (NAD 83)

Degnen Bay, Gabriola Island
Chart 3313, p. 19;
anchor: 49°08.24' N, 123°42.66' W (NAD 83)

Delkatla Inlet Light Chart 3895;
position: (entrance to inlet):
54°00.22' N, 132°08.68' W (NAD 27)

Denise Inlet Chart 3955;
entrance: 54°15.87' N, 130°12.08' W;
anchor: 54°16.63' N, 130°09.40' W (NAD 27)

**Denman Island East Ferry
Landing Light** Chart 3527;
position: (Gravelly Bay):
49°29.68' N, 124°42.41' W (NAD 27)

Denman Island Ferry Landing Light
Chart 3527;
position: (W. side of isl.):
49°32.11' N, 124°49.37' W (NAD 27)

Denman Island Light Chart 3527;
position: (on reef W. side of isl., southward of
Denman Pt): 49°32.30' N, 124°49.72' W
(NAD 27)

Denman Island Public Floats
Chart 3527;
public floats: 49°32.03' N, 124°49.22' W
(NAD 27)

Dent Islands Light Chart 3543;
position: (westerly extremity of Little Dent
Isl.): 50°24.53' N, 125°12.52' W (NAD 83)

Dent Rapids Chart 3543 (inset);
light, Little Dent Island:
50°24.53' N, 125°12.52' W (NAD 27)

Departure Bay Chart 3313, p. 21;
anchor: 49°12.33' N, 123°57.78' W (NAD 83)

**Departure Bay West Cardinal Light
Buoy PW** Chart 3457;
position: 49°11.82' N, 123°56.85' W
(NAD 27)

"Derby Point East" Chart 3535
(Welcome Passage inset);
position: 49°30.53' N, 123°58.69' W
(NAD 27)

Desbrisay Bay, "Big Bay" Chart 3962;
entrance: 52°45. 50' N, 127°59.50' W (NAD 27)

Descanso Bay Light Chart 3458;
position: (on pt W side of bay): 49°10.58' N,
123°52.11' W (NAD 83)

"Descubierta Point Cove" Chart 3664;
anchor: 49°41.50' N, 126°29.25' W (NAD 27)

Deserted Bay Chart 3312;
position: 50°05.30' N, 123°44.60' W
(NAD 27)

Deserters Island Chart 3549;
position: 50°52.65' N, 127°29.28' W (NAD 83)

Desolation Sound Charts 3312, 3538;
entrance: 50°03.50' N, 124°51.50' W
(NAD 27)

Devastation Channel Chart 3743;
south entrance: 53°34.50' N, 128°48.02' W;
north entrance: 53°46.10' N, 128°49.50' W
(NAD 27)

Devastation Island Light Chart 3955;
position: (N. side of isl.):
54°19.37' N, 130°29.12' W (NAD 27)

Devils Hole Chart 3543 (inset);
approximate position:
50°24.23' N, 125°12.37' W (NAD 27)

Devlin Bay Charts 3795, 3724 and 3742;
entrance: 53°04.05' N, 129°35.25' W;
anchor: 53° 03.65' N, 129° 36.50' W
(NAD 27)

Diamond Bay Chart 3537;
position: 50°18.13' N, 125°13.25' W (NAD 27)

Dibuxante Point Light Chart 3475;
position: (W. entrance to Gabriola Pass):
49°07.62' N, 123°42.94' W (NAD 27)

Dillon Bay Chart 3742;
entrance: 53°12.07' N, 129°29.69' W
(NAD 27)

Dillon Rock Light Chart 3549;
position: (on the rock, Sushartie Bay
entrance): 50°51.42' N, 127°51.42' W
(NAD 83)

Dionisio Point Park Chart 3313, p. 16;
anchor: 49°00.85' N, 123°34.48' W (NAD 83)

"Discovery Cove" Chart 3720;
entrance: 52°14.20' N, 128°01.10' W;
anchor: 52°13.78' N, 128°00.32' W (NAD 27)

Discovery Island Light Chart 3424;
position: (extremity of Isl., Haro Strait):
48°25.47' N, 123°13.55' W (NAD 83)

**Discovery Island Marine Park
(Rudlin Bay)** Chart 3313, p. 4;
position: 48°25.27' N, 123°13.77' W (NAD 83)

Discovery Passage Chart 3539;
south entrance: 49°58.50' N, 125°11.00' W
(NAD 83)

Diver Bay Chart 3313, p. 8;
anchor: 48°49.48' N, 123°21.85' W (NAD 83)

Dixie Cove Chart 3682;
anchor: 50°03.2' N, 127°12' W (NAD 27)

Dixon Bay Chart 3648;
anchor: 49°24.20' N, 126°10.15' W) (NAD 27)

Dixon Entrance ODAS Light Buoy
Chart 3802;
position: (NNW of Wiah Pt):
54°22.97' N, 132°25.58' W (NAD 27)

**Dixon Entrance West ODAS Light
Buoy 46205** Chart 3000;
position: 54°10.00' N, 134°20.00' W
(NAD 27)

Doben Island Passage Charts 3920;
anchor (north side of the two islets):
55°26.93' N, 129°46.37' W (NAD 83)

Dock Island Light Chart 3476;
position: (E. end of NE islet of Little Group):
48°40.30' N, 123°21.33' W (NAD 27)

Doctor Bay Chart 3541;
anchor: 50°15.12' N, 124°49.10' W (NAD 27)

Dodd Narrows Chart 3313 (inset), p. 20;
south entrance: 49°08.00' N, 123°48.90' W;
north entrance: 49°08.30' N, 123°49.15' W
(NAD 83)

Dodd Passage Charts 3959, 3963;
west entrance: 54°33.72' N, 130°27.25' W
(NAD 83)

Dodge Cove Chart 3958;
entrance (to range course 228° M):
54°17.56' N, 130°22.00' W (NAD 27)

Dodge Cove Light Buoy D50
Chart 3958;
position: (E. of Dodge Isl.):
54°17.53' N, 130°22.27' W (NAD 83)

Dodge Cove Range Light Chart 3958;
position: (W. end of channel entrance):
54°17.42' N, 130°22.85' W (NAD 83)

Dodge Island Light Buoy D53
Chart 3958;
position: (NE of isl.):
54°17.73' N, 130°22.10' W (NAD 83)

Dodger Channel Chart 3671;
NW anchor: 48°50.21' N, 125°11.96' W;
SE anchor: 48°50.06' N, 125°11.60' W
(NAD 27)

"Dodwell North Cove" Chart 3786;
position: 52°00.40' N, 128°13.35' W
(NAD 27)

"Dodwell South Cove" Chart 3786;
position: 51°59.88' N, 128°13.06' W (NAD 27)

"Dogfish Bay," Kendrick Island
Chart 3313, p. 19;
anchor: 49°07.29' N, 123°41.42' W (NAD 83)

"Dol Cove," Hardy Island Chart 3312;
anchor: 49°43.16' N, 124°12.65' W (NAD 27)

Dolomite Narrows,
"Burnaby Narrows" Chart 3809;
north entrance: 52°21.90' N, 131°21.05' W;
south entrance: 52°21.00' N, 131°20.80' W
(NAD 27)

Dolphin Lagoon Charts 3747, 3927;
entrance: 53°46.68' N, 130°28.23' W (NAD 27)

"Domestic Tranquility Cove"
Chart 3784;
entrance: 51°50.29' N, 128°06.71' W,
anchor: 51°50.41' N, 128°06.70' W (NAD 27)

"Don Peninsula Inlet" Chart 3940;
entrance: 52°21.68' N, 128°12.10' W (NAD 83)

Donald Islets Light Chart 3680;
position: (N side of large islet):
50°13.86' N, 127°48.40' W (NAD 27)

Donald Point Light Chart 3720;
position: (on the pt, Return Channel):
52°17.62' N, 128°05.89' W (NAD 27)

"Donkey Cove" Chart 3891;
entrance: 53°09.10' N, 132°12.12' W;
anchor: 53°09.24' N, 132°12.12' W (NAD 83)

Dorman Bay Chart 3534; (inset)
anchor: 49°22.53' N, 123°19.58' W (NAD 83)

Dorothy Island Light Chart 3743;
position: (on isl.): 53°39.60' N, 128°50.53' W
(NAD 27)

Dory Passage Chart 3746;
east entrance: 53°33.82' N, 130°04.10' W;
west entrance: 53°33.65' N, 130°06.17' W
(NAD 27)

Double Bay, Hanson Island
Chart 3546;
anchor: 50°35.18' N, 126°45.50' W (NAD 83)

"Double Beach Cove" Chart 3649;
anchor: 49°12.02' N, 125°40.70' W (NAD 27)

"Double Eagle Cove" Chart 3552;
anchor: 51°03.94' N, 127°25.20' W (NAD 27)

Double Island Light Chart 3663;
position: (on the isl.):
49°50.67' N, 126°59.80' W (NAD 27)

Double Islets Light Chart 3920;
position: 54°57.75' N, 129°56.82' W (NAD 83)

Douglas Bay Chart 3526;
position: 49°31.26' N, 123°21.38' W (NAD 27)

Douglas Bay, Forward Harbour
Chart 3544;
anchor: 50°28.93' N, 125°45.29' W (NAD 83)

Douglas Channel Charts 3743, 3742;
north entrance: 53°48.00' N 128°52.00' W
(NAD 27)

Doyle Island Light Chart 3549;
position: (SE extremity of isl.):
50°48.34' N, 127°27.54' W (NAD 83)

Draney Inlet Chart 3931;
entrance: 51°28.50' N, 127°34.00' W (NAD 83)

Draney Narrows Chart 3931 (inset);
position (shoal rock):
51°28.42' N, 127°33.77' W (NAD 83)

Drew Harbour Chart 3538;
Rebecca Spit light:
50°06.48' N, 125°11.65' W (NAD 27)

Drew Harbour Light Chart 3538;
position: (near extremity of Rebecca Spit, at
entrance to hbr): 50°06.48' N, 125°11.65' W
(NAD 83)

Dries Inlet Charts 3761, 3927;
entrance: 53°56.25′ N, 130°37.00′ W (NAD 27)

Driver Point Light Chart 3648;
position: (Stewardson Inlet entrance):
49°26.75′ N, 126°15.97′ W (NAD 27)

Drury Inlet Chart 3547;
entrance: 50°53.50′ N, 126°53.50′ W
(NAD 83)

Dryad Point Sector Light Chart 3787;
position: (N. entrance,
Main Passage, Seaforth Channel):
52°11.13′ N, 128°06.60′ W (NAD 27)

Dsulish Bay Chart 3934;
anchor: 51°20.37′ N, 127°40.55′ W (NAD 83)

Du Vernet Point Light Chart 3955;
position: (on drying rock, off pt):
54°18.77′ N, 130°23.97′ W (NAD 27)

Duck Bay Chart 3313, pp. 14, 17;
anchor: 48°53.30′ N, 123°34.67′ W (NAD 83)

Duckers Islands Light Chart 3724;
position: (on easternmost isl.):
52°55.52′ N, 129°11.48′ W (NAD unknown)

Dudevoir Passage Chart 3963;
west entrance: 54°37.86′ N, 130°26.71′ W;
east entrance: 54°38.29′ N, 130°26.93′ W
(NAD 83)

Duff Islet Light Chart 3576;
position: (on islet, entrance to Fife Sound):
50°45.41′ N, 126°43.30′ W (NAD unknown)

Duffin Cove Chart 3685;
anchor: 49°09.08′ N, 125°54.72′ W (NAD 83)

Dugout Rocks Light Chart 3934;
position: (eastern side of approach to Fitz
Hugh Sound): 51°22.03′ N, 127°48.39′ W
(NAD 83)

Duncan Bay
Charts 3955 (inset), 3957, 3959;
west entrance: 54°21.20′ N, 130°30.50′ W;
anchor (0.12 mile south Hecate Rock):
54°20.33′ N, 130°28.20′ W (NAD 83)

"Duncanby Cove" Chart 3934;
anchor: 51°24.42′ N, 127°40.17′ W (NAD 83)

Duncanby Landing Chart 3934;
floats: 51°24.35′ N, 127°38.75′ W (NAD 83)

Dundas Island Light Chart 3959;
position: (entrance to Edith Hbr):
54°27.58′ N, 130°56.75′ W (NAD 83)

Dundas Islands Chart 3959;
west entrance (Hudson Bay Passage):
54°25.50′ N, 130°58.50′ W;
east entrance: 54°33.00′ N, 130°45.00′ W
(NAD 83)

Dundas Point Sector Light Chart 3955;
position: (NW of pt):
54°19.43′ N, 130°24.93′ W (NAD 27)

Dundivan Inlet Chart 3787;
entrance: 52°13.70′ N, 128°15.30′ W
(NAD 27)

Dunn Bay Chart 3720;
position: 52°10.35′ N, 127°58.10′ W (NAD 27)

Dunsmuir Point Light Chart 3668;
position: (W. side of Alberni Inlet,
opposite mouth of China Creek):
49°09.24′ N, 124°48.50′ W (NAD 83)

Dupont Island Light Chart 3724;
position: 52°56.39′ N, 129°26.17′ W
(NAD unknown)

Dyer Cove Chart 3711 (inset);
entrance: 52°11.60′ N, 128°28.42′ W;
anchor: 52°11.08′ N, 128°28.16′ W (NAD 27)

Eagle Bay Chart 3743;
entrance: 53°49.20' N, 128°42.80' W;
anchor: 53° 48.11' N, 128° 42.37' W
(NAD 27)

"Eagle Cove," Stagoo Creek
Chart 3933;
anchor: 55°17.65' N, 129°45.02' W (NAD 27)

Earl Ledge Chart 3544;
light: 50°24.66' N, 125°55.25' W (NAD 83)

Earl Ledge Light Chart 3544;
position: (seaward end of ledge):
50°24.66' N, 125°55.25' W (NAD 83)

Earls Cove Charts 3512, 3514;
ferry terminal light:
49°45.20' N, 124°00.45' W (NAD 27)

Earls Cove Light Chart 3514;
position: 49°45.20' N, 124°00.45' W
(NAD 83)

East Bear Bight Chart 3544;
anchor: 50°21.73' N, 125°38.91' W (NAD 83)

East Copper Island Light Chart 3809;
position: 52°21.43' N, 131°10.33' W
(NAD 27)

East Cove Chart 3679;
anchor: 50°29.83' N, 127°50.20' W (NAD 83)

"East Cove," Dsulish Bay Chart 3934;
anchor: 51°20.39' N, 127°40.06' W (NAD 83)

East Cracroft Island Range Light
Chart 3564;
position: (northeastern side of isl.):
50°34.78' N, 126°14.47' W (NAD 83)

East Inlet, Inner Basin Chart 3772;
entrance: 53°40.33' N, 129°42.96' W;
anchor (inner basin):
53°42.87' N, 129°43.51' W (NAD 27)

East Kinahan Island Light Chart 3958;
position: (NE extremity of isl.):
54°12.83' N, 130°23.62' W (NAD 83)

East Narrows, Skidegate Channel
Chart 3891 (inset);
east entrance: 53°08.72' N, 132°13.60' W;
west entrance: 53°09.26' N, 132°18.02' W
(NAD 83)

Echo Bay Chart 3515;
float: 50°44.60' N, 126°29.85' W (NAD 83)

Echo Cove Chart 3920;
anchor: 54°55.67' N, 129°57.00' W (NAD 83)

Echo Harbour Chart 3807;
entrance: 52°42.10' N, 131°45.80' W;
anchor (harbor): 52°41.51' N, 131°45.72' W
(NAD 27)

Eclipse Narrows Chart 3552 (inset);
entrance: 51°04.13' N, 126°45.48' W (NAD 27)

Ecoole Charts 3670, 3668;
anchor: 48°58.03' N, 125°03.41' W (NAD 83)

Ecstall Island Light Chart 3717;
position: (N. extremity of Ecstall Isl.):
54°09.75' N, 129°57.33' W (NAD 83)

Edith Cove Chart 3555 (inset);
anchor: 50°30.22' N, 125°36.26' W (NAD 27)

Edith Harbour Chart 3959;
entrance: 54°27.60' N, 130°56.75' W;
mooring buoys: 54°28.03' N, 130°56.98' W
(NAD 83)

Edward Channel Chart 3784;
south entrance: 51°45.00' N, 128°04.77' W;
north entrance: 51°47.03' N, 128°04.40' W
(NAD 27)

Edward Point Light Chart 3729;
position: (on pt): 52°26.38' N, 127°16.30' W
(NAD 27)

Edye Passage Chart 3956;
west entrance: 54°03.30' N, 130°39.75' W;
east entrance: 54°04.20' N, 130°33.20' W
(NAD 83)

Effingham Bay Chart 3670;
anchor: 48°52.45' N, 125°18.23' W (NAD 83)

Egg Island Chart 3550, 3934;
light: 51°14.90' N, 127°49.93' W;
anchor (tiny east cove):
51°14.93' N, 127°49.85' W (NAD 83)

Egg Island Light Chart 3550;
position: (on summit of isl.):
51°14.91' N, 127°49.93' W (NAD 83)

Egmont, Secret Bay Chart 3312;
public float: 49°45.03 N, 123°55.67' W
(NAD 27)

Ehatisaht Chart 3663;
entrance: 49°52.4' N, 126°52.0' W (NAD 27)

Ehatishat Light Chart 3663;
position: (on pt, E. of Ehatishat):
49°52.88' N, 126°49.48' W (NAD 27)

Ekins Point Light Chart 3526;
position: (on pt): 49°32.12' N, 123°22.75' W
(NAD 83)

Elcho Harbour Chart 3781;
entrance: 52°22.40' N, 127°28.90' W;
anchor: 52°23.83' N, 127°31.8' W (NAD 27)

Eliot Passage Chart 3546;
north entrance: 50°37.80' N, 126°35.00' W
(NAD 83)

Eliza Bay and Lucy Bay Chart 3934;
east entrance: 51°19.05' N, 127°44.25' W
(NAD 83)

Elizabeth Lagoon Chart 3921;
position (Rapids):
51°39.23' N, 127°45.85' W (NAD 83)

Elk Bay Chart 3539;
position: 50°16.90' N, 126°26.10' W (NAD 83)

Ellen Bay Chart 3313, p. 8;
anchor: 48°49.20' N, 123°22.60' W (NAD 83)

Ellerslie Bay Chart 3940;
south entrance: 52°31.00' N, 128°01.00' W
(NAD 83)

"Ellerslie Bay, East Anchorage"
Chart 3940;
entrance: 52°30.85' N, 128°00.50' W;
anchor (north shore):
52°39.30' N, 127°59.45' W (NAD 83)

Ellerslie Lagoon, First Narrows
Chart 3940;
entrance: 52°31.78' N, 128°00.96' W;
anchor (northwest of the falls):
52°31.71' N, 127°59.75' W (NAD 83)

Ellis Bay Chart 3552
entrance: 51°03.08' N, 127°24.22' W;
anchor: 51°02.92' N, 127°25.05' W (NAD 27)

Embley Lagoon Chart 3547;
position: 50°56.97' N, 126°52.14' W (NAD 83)

Emilia Island Light Chart 3743;
position: (SE end of isl.):
53°45.48' N, 128°58.37' W (NAD 27)

Emily Bay Chart 3940;
entrance: 52°23.50' N, 128°00.00' W;
anchor: 52°23.53' N, 128°00.83' W (NAD 83)

Emily Carr Inlet
Charts 3719 (inset), 3737;
entrance (lagoon):
52°55.65'N, 129°08.84' W;
cove (east entrance):
52°55.375' N, 129°08.835' W;
anchor (northwest cove):
52°55.39' N, 129°09.38' W (NAD 27)

**Emsley Cove ("Old Town"
or "Bish Bay")** Chart 3743;
position: 53°54.04' N, 128°46.55' W (NAD 27)

"End of the World Inlet" Chart 3787;
entrance: 52°05.00' N, 128°14.25' W
(NAD 27)

**Enfield Rock Light and
Bell Buoy D76** Chart 3957;
position: 54°18.38' N, 130°32.60' W
(NAD 83)

**English Bay Anchorage
East Light** Chart 3481;
position: (seaward end of breakwater):
49°16.53' N, 123°11.22' W (NAD 27)

English Bay Chart 3481;
Jericho public pier position:
49°16.65' N, 123°12.00' W (NAD 27);
Jericho Beach anchor:
49°16.53' N, 123°11.70' W (NAD 27);
Royal Vancouver Yacht Club light on breakwater: 49°16.53' N, 123°11.22' W (NAD 27)

English Bay Light Buoy Q41
Chart 3493;
position: 49°17.12' N, 123°08.85' W
(NAD 83)

Enterprise Reef Light Chart 3473;
position: (on western rock of reef):
48°50.71' N, 123°20.81' W (NAD 27)

Entrance Inlet Chart 3670;
anchor: 49°00.14' N, 125°17.70' W (NAD 83)

Entrance Island Light Chart 3458;
position: (on isl., northern approach to
Nanaimo): 49°12.56' N, 123°48.41' W
(NAD 83)

Epsom Point Light Chart 3311;
position: (on extremity of pt):
49°30.30' N, 124°00.90' W (NAD 83)

Equis Beach and Sechart
Chart: 3670, 3671;
anchor: 48°57.89' N, 125°17.57' W (NAD 83)

Esperanza Chart 3663;
fuel dock: 49°52.33' N, 126°44.41' W
(NAD 27)

Esperanza Inlet Chart 3663;
entrance buoy "MD":
49°47.12' N, 127°02.80' W,
Middle Reef light buoy "M41":
49°48.10' N, 127°02.30' W (NAD 27)

**Esperanza Inlet Light and
Whistle Buoy MD** Chart 3663;
position: 49°47.12' N, 127°02.80' W (NAD 27)

Espinosa Inlet Chart 3663;
entrance: 49°52' N, 126°55' W (NAD 27)

Estero Basin Chart 3543;
'The Cut": 50°30.15' N, 125°14.93' W (NAD 27)

Estevan Point Chart 3640;
light: 49°23.00' N, 126°32.53' W (NAD 27)

Estevan Point Light Chart 3640;
position: (SW extremity of pt at Hole-in-the-
Wall): 49°23.00' N, 126°32.53' W (NAD 27)

Estevan Sound Charts 3724, 3742;
south entrance: 52°57.00' N, 129°23.00' W;
north entrance: 53°09.50' N, 129°37.50' W
(NAD 27)

Ethel Cove Chart 3931;
entrance: 51°20.06' N, 127°31.66' W;
anchor: 51°20.26' N, 127°31.32' W (NAD 83)

Eucott Bay Chart 3729;
anchor: 52°27.25' N, 127°18.95' W (NAD 27)

"Europa Bay" Chart 3743;
entrance: 53°26.75' N, 128°33.60' W;
position: 53°28.40' N, 128°41.00' W (NAD 27)

Europa Point Light Chart 3745;
position: (on pt): 53°25.77' N, 128°32.60' W
(NAD 27)

Evans Bay / Bird Cove
Charts 3539, 3312;
Bird Cove entrance:
50°11.85' N, 125°05.20' W;
north arm entrance:
50°12.80' N, 125°04.40' W (NAD 83)

Evans Inlet Chart 3785;
entrance: 52°06.10' N, 127°52.70' W (NAD 27)

Evening Cove Chart 3475 (inset);
anchor: 48°59.27' N, 123°46.16' W (NAD 27)

"Evinrude Inlet Bight" Chart 3719
(inset);
anchor: 52°47.32' N, 129°05.87' W (NAD 27)

Evinrude Inlet Chart 3719 (inset);
entrance: 52°47.45' N, 129°06.20' W;
log tie-up: 52°48.17' N, 129°04.85' W
(NAD 27)

Evinrude Passage Chart 3746;
west entrance: 53°32.23' N, 129°59.03' W;
east entrance: 53°32.90' N, 129°57.10' W
(NAD 27)

Experiment Bight Chart 3624;
anchor: 50°46.83' N, 128°24.50' W (NAD 27)

Exposed Inlet Chart 3772;
entrance: 53°40.00' N, 129°42.90' W;
position: 53°39.80' N, 129°42.50' W (NAD 27)

Fair Harbour Approach Light
Chart 3682;
position: 50°04.49' N, 127°09.29' W
(NAD 27)

Fair Harbour Chart 3682;
position: 50°04' N, 127°07' W (NAD 27)

Fair Harbour Light Chart 3682;
position: 50°04.13' N, 127°08.55' W
(NAD 27)

"Fallen Human Bay" Chart 3955;
entrance: 54°19.00' N, 130°22.70' W;
anchor: 54°19.53' N, 130°23.88' W (NAD 27)

False Bay ("Orchard Bay") Chart 3536;
anchor: 49°29.88' N, 124°21.32' W (NAD 27)

False Bay Light Chart 3536;
position: (On Prowse Pt at S. side of entrance
to bay): 49°29.25' N, 124°21.65' W (NAD 27)

False Cove Charts 3546, 3515;
position: 50°43.95' N, 126°32.90' W
(NAD 83)

False Creek Chart 3493; red buoy "Q52"
position: 49°16.90' N, 123°08.90 W
(NAD 83);
sector light: 49°16.56' N, 123°08.10' W
(NAD 83)

False Creek Sector Light Chart 3493;
position: (N. pier of Burrard Bridge):
49°16.56' N, 123°08.10' W (NAD 83)

"False Gay Passage" Chart 3683
north entrance: 50°05.65' N, 127°31.80' W
(NAD 27)

False Lagoon ("Johnson Lagoon")
Chart 3536 (inset);
position: 49°29.00' N, 124°21.00' W (NAD 27)

False Narrows Chart 3475; 3313,
p. 20 (inset)
east entrance: 49°07.42' N, 123°45.67' W
west entrance: 49°08.23' N, 123°44.53' W
(NAD 27)

Fancy Cove Chart 3785;
entrance: 52°04.00' N, 128°01.20' W;
anchor: 52°03.68' N, 128°00.63' W (NAD 27)

Fane Island Light Chart 3477;
position: 48°48.43' N, 123°16.00' W
(NAD 27)

Fannie Cove Chart 3785;
position: 52°02.92' N, 128°04.02' W (NAD 27)

Fanny Bay Chart 3527;
float: 49°30.46' N, 124°49.53' W (NAD 27)

Fanny Island Light Chart 3544;
position: 50°27.21' N, 125°59.57' W
(NAD 83)

Farewell Harbour Chart 3546;
anchor: 50°36.08' N, 126°40.36' W (NAD 83)

Farewell Point Light Chart 3785;
position: (E. side of pt):
52°07.57' N, 127°53.33' W (NAD 27)

Farrant Island Lagoon Chart 3772;
entrance: 53°23.22' N, 129°26.72' W;
anchor: 53°22.76' N, 129°26.03' W (NAD 27)

Farrer Cove Chart 3495;
anchor: 49°20.15' N, 122°53.32' W (NAD 83)

"Farwest Cove" Chart 3959;
position (Farwest Point):
54°25.50' N, 130°50.00' W;
entrance (cove): 54°25.94' N, 130°48.70' W
(NAD 83)

Fegan Islets Light Chart 3512;
position: (W. entrance to Sabine Channel):
49°31.95' N, 124°22.90' W (NAD 27)

Felice Island Light Buoy Y3
Chart 3685;
position: (S. of isl.):
49°08.77' N, 125°55.38' W (NAD 83)

**Ferguson Point West Cardinal
Light Buoy QC** Chart 3493;
position: 49°18.22' N, 123°09.93' W
(NAD 83)

Fern Passage Chart 3958;
entrance: 54°20.15' N, 130°16.70' W (NAD 27)

Fernie Island Light Chart 3476;
position: (on isl. SE of isl.):
48°40.73' N, 123°23.41' W (NAD 27)

Fiddle Reef Sector Light Chart 3424;
position: (on reef): 48°25.76' N,
123°17.03' W (NAD 83)

Fife Sound Charts 3547, 3515;
west entrance: 50°45.65' N, 126°43.80' W
(NAD 83)

Fifer Bay Charts 3921, 3934;
entrance: 51°35.80' N, 127°49.90' W;
anchor: 51°35.44' N, 127°49.51' W (NAD 83)

Fifer Cove Chart 3737;
entrance: 52°52.30' N, 128°41.60' W;
anchor: 52°51.67' N, 128°40.77' W (NAD 27)

Fin Rock Light Chart 3742;
position: 53°13.97' N, 129°21.55' W (NAD 27)

Finis Nook Chart 3931;
entrance: 51°20.85' N, 127°30.07' W;
anchor: 51°20.47' N, 127°30.41' W (NAD 83)

Finisterre Island Light Chart 3526;
position: (N. extremity of isl.):
49°25.10' N, 123°18.43' W (NAD 83)

Finlayson Channel Charts 3728, 3734
south entrance: 52°24.65' N, 128°29.30' W
northwest entrance:
52°48.75' N, 128°26.35' W (NAD 27)

Finn Bay Chart 3934;
entrance: 51°30.42' N, 127°43.36' W;
position: 51°30.07' N, 127°44.10' W (NAD 83)

Finn Cove Charts 3311, 3538;
entrance: 49°59.00' N, 124°46.00' W
(NAD 83)

"First Narrows Cove" Chart 3940,
anchor: 52°21.78' N, 128°00.28' W (NAD 83)

First Narrows Light Chart 3493;
position: (entrance to First Narrows):
49°19.15' N, 123°08.70' W (NAD 83)

Fisgard Island Light Buoy V17
Chart 3419;
position: (off island):
48°25.84' N, 123°26.60' W (NAD 83)

Fisgard Island Light Buoy V17
Chart 3419;
position: (on island):
48°26.53' N, 123°26.26' W (NAD 83)

Fisgard Sector Light Chart 3419;
position: (on Fisgard Isl., W. side of entrance
to Esquimalt Hbr):
48°25.83' N, 123°26.77' W (NAD 83)

Fish Egg Inlet Chart 3921;
note the entrance waypoints listed below
(NAD 83)

Fish Trap Bay Chart 3921;
position: 51°37.45' N, 127°41.74' W (NAD 83)

"Fish Weir Cove," Spiller Channel
Chart 3940;
entrance: 52°28.05' N, 128°02.20' W;
anchor (north side):
52°28.15' N, 128°01.98' W (NAD 83)

Fisher Channel Chart 3785;
south entrance: 51°55.50' N, 127°56.25' W;
north entrance: 52°15.70' N, 127° 46.40' W
(NAD 27)

Fisherman Bay Charts 3598 or 3624
(same scale, better detail);
anchor: 50°47.75' N, 127°53.85' W (NAD 27)

Fisherman Cove Chart 3742;
position: 53°19.55' N, 129°16.70' W (NAD 27)

Fishermans Cove Light Chart 3534;
position: 49°21.27' N, 123°16.52' W
(NAD 83)

Fishermans Harbour Chart 3493;
position: 49°16.43' N, 123°08.17' W (NAD 83)

Fishhook Bay Chart 3931;
anchor: 51°27.69' N, 127°33.11' W (NAD 83)

Fishtrap Bay Chart 3743;
position: 53°33.06' N, 129°01.24' W (NAD 27)

Fitz Hugh Sound Chart 3934;
south entrance: 51°23.50' N, 127°50.00' W;
north entrance: 51°55.50' N, 127°56.25' W
(NAD 83)

"Five Meter Hole" Chart 3940;
entrance: 52°22.35' N, 128°12.83' W;
anchor: 52°22.53' N, 128°12.81' W (NAD 83)

"Five Window Cove" Chart 3934;
entrance: 51°28.73' N, 127°40.75' W;
anchor: 51°29.08' N, 127°39.95' W (NAD 83)

Flagpole Point Light Chart 3730;
position: (on pt): 52°20.93' N, 126°55.60' W
(NAD 27)

Flatrock Island Light Chart 3855;
position: (near highest point on isl.):
52°06.47' N, 131°10.03' W (NAD 27)

"Float Camp Cove" Chart 3934;
position: 51°31.52' N, 127°40.28' W (NAD 83)

Flora Islet Light Chart 3527;
position: (N. side of islet, off St. John Pt):
49°31.05' N, 124°34.50' W (NAD 27)

Florence Cove Charts 3537, 3539;
anchor: 50°18.63' N, 125°09.88' W (NAD 27)

Flowery Islet Light Chart 3890;
position: (highest point of islet):
53°13.35' N, 132°00.34' W (NAD 27)

Fly Basin Charts 3934, 3931;
entrance: 51°16.65' N, 127°36.60' W;
anchor: 51°16.30' N, 127°36.00' W (NAD 83)

Foch Lagoon, Drumlummon Bay
Chart 3743;
lagoon entrance: 53°45.85' N, 129°01.45' W
(NAD 27)

Fog Rocks Light Chart 3785;
position: (on largest rock, Fisher Channel):
51°58.35' N, 127°55.03' W (NAD 27)

Folger Island Light Chart 3671;
position: (W. pt of isl.):
48°49.79' N, 125°14.96' W (NAD 27)

Forbes Bay Chart 3541;
anchor "South Nook":
50°14.09' N, 124°35.97' W (NAD 27)

Forbes Island Light Chart 3670;
position: (S. side of Isl.):
48°56.95' N, 125°24.54' W (NAD 83)

Ford Cove Charts 3527, 3513;
public float: 49°29.82' N, 124°40.53' W
(NAD 27)

Fords Cove Chart 3933;
entrance: 55°37.78' N, 130°05.80' W;
anchor (small boats):
55°37.67' N, 130°05.65' W (NAD 27)

Forit Bay Chart 3720;
anchor: 52°10.38' N, 127°54.63' W (NAD 27)

Forrest Island Light Chart 3441;
position: 48°39.31' N, 123°19.32' W (NAD 27)

Fort Point Light Chart 3920;
position: 54°49.10' N, 129°54.95' W (NAD 83)

Fort Rupert Chart 3546;
position: 50°42.05' N, 127°24.10' W (NAD 83)

Forward Bay Chart 3545;
anchor: 50°31.10' N, 126°24.70' W (NAD 83)

Forward Inlet Chart 3686
entrance: 50°27.90' N, 128°01.20' W (NAD 83)

Fougner Bay Charts 3785 or 3729;
entrance: 51°54.60' N, 127°51.60' W;
anchor (outer anchorage):
51°54.40' N, 127°51.20' W;
anchor (inner cove):
51°54.28' N, 127°50.60' W (NAD 27)

Fox Islands Chart 3550;
east entrance: 51°05.08' N, 127°35.40' W;
west entrance: 51°04.55' N, 127°36.70' W
(NAD 83)

Frances Bay Chart 3541, 3312;
anchor: 50°21.00' N, 125°02.48' W
(NAD 83)

Francis Island Light Chart 3646;
position: (Ucluelet Hbr entrance):
48°55.33' N, 125°31.27' W (NAD 83)

Francis Point Light Chart 3535;
position: (on pt): 49°36.22' N, 124°03.45' W
(NAD 27)

Franklin River Light Chart 3668;
position: (at Sproat Narrows, on edge of
shoal off mouth of river):
49°06.22' N, 124°49.25' W (NAD 83)

"Franks Bay" ("Jane Bay")
Chart 3671, 3668;
anchor South Cove: 49°00.25' N, 125°08.77' W
(NAD 27)

Fraser River, Main Channel
Chart 3490;
Sand Heads light:
49°06.19' N, 123°18.57' W (NAD 83)

Fraser River, Middle Arm Chart 3491;
Middle Arm swing bridge position:
49°11.60' N, 123°08.10' W (NAD 27)

Fraser River, North Arm Chart 3481;
North Arm jetty light:
49°15.45' N, 123°16.70' W (NAD 27)

Frederick Arm Chart 3543
entrance: 50°27.45′ N, 125°17.00′ W (NAD 27)

Frederick Bay Chart 3552;
position: 51°02.20′ N, 127°14.82′ W (NAD 27)

Frederick Island Light Chart 3868;
position: (on Hope Point):
53°56.30′ N, 133°11.80′ W (NAD 27)

Frederick Point Light Chart 3958;
position: (at edge of bar):
54°15.50′ N, 130°21.50′ W (NAD 83)

Frederick Sound Chart 3552 (inset);
entrance: 51°04.13′ N, 126°45.48′ W;
anchor (basin) 50°59.27′ N, 126°44.48′ W
(NAD 27)

Frederiksen Point Charts 3598 or 3624;
way-point: 50°49.50′ N, 128°21.50′ W
(NAD 27)

Freeman Passage Charts 3761, 3927;
west entrance: 53°49.40′ N, 130°39.75′ W;
east entrance: 53°51.10′ N, 130°34.20′ W
(NAD 27)

**Freeman Passage Entrance Light
and Bell Buoy E98** Chart 3761;
position: (NW of Joachim Pt):
53°49.43′ N, 130°39.38′ W (NAD 27)

Freeman Passage Light Chart 3761;
position: (SW isl. of group):
53°49.83′ N, 130°37.65′ W (NAD 27)

Freeman Passage Mooring Buoys
Charts 3761, 3927;
position: 53°50.49′ N, 130°38.59′ W (NAD 27)

Freeman Point Light Chart 3711;
position: (S. extremity of Cone Isl.):
52°33.19′ N, 128°29.30′ W (NAD 27)

Freke Anchorage Charts 3559, 3312, 3538
position: 49°58.20′ N, 124°40.90′ W (NAD 27)

French Creek Chart 3512;
west breakwater light:
49°21.12′ N, 124°21.27′ W (NAD 27)

French Creek Light Chart 3512;
position: (near outer end of W. breakwater):
49°21.12′ N, 124°21.27′ W (NAD 27)

Frenchmans Cove, Halfmoon Bay
Chart 3535
(Welcome Passage inset);
anchor: 49°30.46′ N, 123°56.83′ W (NAD 27)

"Freshwater Cove," Hoya Passage
Charts 3807, 3808;
entrance: 52°40.10′ N, 131°43.30′ W;
anchor: 52°39.93′ N, 131°43.60′ W (NAD 27)

Friendly Cove Chart 3664;
anchor: 49°35.67′ N, 126°36.93′ W (NAD 27)

"Friendly Dolphin Cove" Chart 3648;
anchor: 49°23.88′ N, 126°05.48′ W (NAD 27)

Frigon Islets Light Chart 3681;
position: (SW pt of most westerly
islet of Frigon Isl. Group):
50°25.07′ N, 127°29.64′ W (NAD 83)

Frypan Bay Chart 3934;
entrance: 51°29.64′ N, 127°42.02′ W;
anchor: 51°29.65′ N, 127°42.40′ W (NAD 83)

Fulford Harbour Light Chart 3470;
position: (on ferry landing slip):
48°46.19′ N, 123°27.02′ W (NAD unknown)

**Fulford Harbour,
Saltspring Island** Chart 3313, p. 8;
ferry landing light: 48°46.19′ N, 123°27.02′ W;
anchor: 48°46.16′ N, 123°27.32′ W (NAD 83)

**Fulford Reef North Cardinal
Light Buoy VK** Chart 3424;
position: 48°26.88′ N, 123°14.38′ W
(NAD 83)

"Fury Bay" Chart 3737;
position: 52°40.02' N, 129°00.75' W (NAD 27)

"Fury Cove," Schooner Retreat
Chart 3934;
anchor: 51°29.25' N, 127°45.65' W (NAD 83)

Schooner Retreat
entrance: 51°28.57' N, 127°45.73' W
(NAD 83)
Frigate Bay entrance
(Safe Entrance):
51°27.78' N, 127°45.00' W;
anchor: 51°28.40' N, 127°43.75' W (NAD 83)
Secure Anchorage
anchor: 51°28.48' N, 127°44.42' W
(NAD 83)
Exposed Anchorage
position (turn for Fury Cove):
51°28.90' N, 127°45.15' W (NAD 83)
Rocky Bay
position: 51°28.83' N, 127°44.26' W
(NAD 83)

Gabriola Passage East Light
Chart 3475;
position: (off NE side, Valdes Isl.):
49°07.50' N, 123°41.22' W (NAD 27)

**Gabriola Reefs Bifurcation Light
Buoy UM** Chart 3475;
position: (S. extremity of reef):
49°07.68' N, 123°39.27' W (NAD 27)

Gale Passage Chart 3787;
entrance: 52°14.55' N, 128°22.30' W (NAD 27)

Gale Passage Chart 3787;
north entrance: 52°14.50' N, 128°22.30' W
(NAD 27)

**"Gale Passage Landlocked South
Cove"** Chart 3787;
position: 52°10.95' N, 128°23.00' W (NAD 27)

Galiano Bay Chart 3664;
anchor: 49°42.30' N, 126°27.90' W (NAD 27)

Galiano Light Chart 3473;
position: (on drying rock):
48°51.73' N, 123°20.85' W (NAD 27)

Galley Bay Charts 3559, 3312, 3538;
anchor "West Nook":
50°04.18' N; 124°47.50' W;
anchor "East Island":
50°04.34' N, 124°46.64' W (NAD 27)

Gallows Point Light Chart 3457;
position: (on the pt):
49°10.22' N, 123°54.99' W (NAD 27)

Ganges Harbour Chart 3313, p. 10;
anchor: 48°51.35' N, 123°29.75' W;
public wharf: 48°51.26' W, 123°29.91' W
(NAD 83)

Ganges Harbour Light Chart 3478;
position: (on Second Sister Isl.):
48°50.22' N, 123°27.13' W (NAD 83)

Garcin Rocks Light Chart 3855;
position: (middle of rocks):
52°12.52' N, 130°57.92' W (NAD 27)

Garden Bay Chart 3535 (inset);
anchor: 49°37.85' N, 124°01.28' W (NAD 27)

Garden Beach (Beach Gardens Marina)
Charts 3311, sheet 4;
south breakwater light:
49° 48.02'N, 124°31.07'W (NAD 83)

"Garden Point Bight" Chart 3663;
anchor: 49°50.85' N, 126°53.54' W (NAD 27)

Gardner Canal Charts 3743, 3745;
entrance: 53°34.50' N, 128°48.00' W (NAD 27)

Garry Point Light Chart 3490;
position: (at pt): 49°07.47' N, 123°11.73' W
(NAD 83)

Garry Point West Light Chart 3490;
position: (W. of pt):
49°07.50' N, 123°11.92' W (NAD 83)

"Gasboat Cove" Charts 3761, 3927;
entrance: 53°49.86' N, 130°25.26' W;
anchor: 53°50.22' N, 130°24.84' W (NAD 27)

Gasboat Passage Charts 3761, 3927;
east entrance: 53°49.60' N, 130°20.30' W;
west entrance: 53°49.60' N, 130°25.00' W
(NAD 27)

Gay Passage Chart 3683;
north entrance: 50°06.63' N, 127°31.70' W
(NAD 27)

Gayward Rock Light Chart 3651;
position: 50°01.27' N, 127°23.37' W
(NAD 83)

"Gee Whiz Nook" Chart 3921;
anchor: 51°36.37' N, 127°43.82' W (NAD 83)

Genn Islands Light Chart 3717;
position: (NW extremity of isl.):
54°05.86' N, 130°17.48' W (NAD 83)

Genoa Bay Chart 3313, p. 14;
anchor: 48°45.86' N, 123°35.87' W (NAD 83)

Geodetic Cove Chart 3742;
entrance: 53°06.10' N, 129°37.35' W;
anchor: 53°05.52' N, 129°37.68' W (NAD 27)

George Point Light Chart 3895;
position: (on pt, entrance to Naden Hbr):
54°02.49' N, 132°33.97' W (NAD 27)

Georgeson Passage Chart 3313, p. 12;
south entrance: 48°49.15' N, 123°13.10' W
(NAD 83)

**Georgia Rock Light and
Bell Buoy D43** Chart 3958;
position: (E. of rock):
54°13.26' N, 130°21.62' W (NAD 83)

Gerald Island Cove Chart 3459;
49°18.78' N, 124°09.79' W (NAD 27)

Gerrans Bay ("Whiskey Slough")
Chart 3535 (inset);
position: 49°37.25' N, 124°02.40' W
(NAD 27)

Gertrude Point Light Chart 3743;
position: (on pt, Douglas Channel):
53°37.93' N, 129°13.70' W (NAD 27)

Gibson Cove Chart 3649;
entrance: 49°20' N, 125°58' W (NAD 27)

**Gibson Island Light and
Bell Buoy D6** Chart 3773;
position: 53°55.00' N, 130°09.85' W
(NAD 27)

Gibsons Chart 3534;
anchor: 49°24.14' N, 123°30.10' W (NAD 83)

**Gibsons Landing Breakwater
South Light** Chart 3534;
position: 49°23.97' N, 123°30.20' W (NAD 83)

Gibsons Landing Light Chart 3534;
position: (outer end of breakwater):
49°23.99' N, 123°30.17' W (NAD 83)

Gibsons Landing Rock Light
Chart 3534;
position: (on rock):
49°23.93' N, 123°29.83' W (NAD 83)

Gildersleeve Bay, West Cove
Chart 3921;
position (west cove):
51°36.07' N, 127°46.53' W;
entrance (lagoon):
51°35.98' N, 127°46.24' W (NAD 83)

"Gill Net Cove" Charts 3934, 3931;
anchor: 51°18.05' N, 127°34.17' W (NAD 83)

Gillard Islands Light Chart 3543;
position: (NE extremity of isl., Gillard
Passage): 50°23.49' N, 125°09.26' W
(NAD 83)

Gillard Passage Chart 3543 (inset);
center passage position:
50°23.58′ N, 125°09.42′ W (NAD 27)

Gillatt Island Light Chart 3890;
position: (western end of isl.):
53°14.69′ N, 131°53.87′ W (NAD 27)

Gillen Harbour Chart 3723 (inset);
entrance (south channel):
52°57.60′ N, 129°35.67′ W;
anchor: 52°58.99′ N, 129°35.48′ W
(NAD 27)

Gillespie Channel Chart 3795;
east entrance
53°03.60′ N, 129°36.22′ W (NAD 27)

Gilttoyees Inlet Chart 3743;
entrance: 53°46.20′ N, 128°57.75′ W;
anchor (outer anchorage):
53°47.00′ N, 128°57.00′ W (NAD 27)

"Glacier Bay and Cove" Chart 3933;
entrance: 55°48.95′ N, 130°07.60′ W;
anchor (east cove):
55°49.38′ N, 130°06.93′ W;
anchor (west bay):
55°49.45′ N, 130°07.56′ W (NAD 27)

Glaholm Islet Light Chart 3720;
position: (off NW end of Dearth Isl.):
52°15.82′ N, 128°12.74′ W (NAD 27)

Glendale Cove Chart 3515;
position: 50°40.04′ N, 125°43.85′ W
(NAD 83)

Glenthorne Passage Chart 3313, p. 10;
anchor: 49°49.27′ N, 123°23.19′ W (NAD 83)

Goat Cove Chart 3738;
entrance: 52°47.10′ N, 128°24.60′ W;
anchor (inner cove):
52°46.32′ N, 128°23.34′ W (NAD 27)

Goat Harbour Chart 3740, 3742;
entrance: 53°21.20′ N, 128°53.10′ W;
head of bay: 53°21.68′ N, 128°51.24′ W
(NAD 27)

Gobeil Bay, "Mud Bay" Chart 3743;
entrance: 53°52.00′ N, 128°40.65′ W;
anchor: 53°52.63′ N, 128°40.25′ W (NAD 27)

God's Pocket Chart 3549;
anchor buoys: 50°50.42′ N, 127°35.62′ W
(NAD 83)

"God's Pocket" Chart 3649;
anchor: 49°12.82′ N, 125°53.10′ W (NAD 27)

Godkin Point Light Chart 3549;
position: 50°53.75′ N, 127°55.68′ W (NAD 83)

Gold River Chart 3664;
float: 49°40.73′ N, 126°06.95′ W (NAD 27)

Gold River Light Chart 3665;
position: (W. side of entrance to Gold River):
49°40.53′ N, 126°07.57′ W (NAD 27)

Goldstream Harbour Chart 3784;
entrance: 51°43.77′ N, 127°59.50′ W
anchor: 51°43.68′ N, 128°00.16′ W (NAD 27)

Goletas Channel Chart 3549;
south entrance: 50°47.00′ N, 127°29.00′ W
(NAD 83)

Gomer Island Chart 3543;
position: 50°27.50′ N, 125°16.10′ W (NAD 27)

Good Hope Chart 3932;
position: 51°34.25′ N, 127°30.95′ W (NAD 83)

Gooding Cove
Charts 3686, 3679 and 3624;
anchor: 50°23.95′ N, 127°57.18′ W (NAD 83)

Goodlad Bay Chart 3786;
entrance: 51°52.30′ N, 128°09.20′ W;
anchor: 51°52.65′ N, 128°09.62′ W (NAD 27)

Goose Bay Chart 3934;
position: 51°22.47′ N, 127°40.09′ W (NAD 83)

Goose Group Chart 3786;
entrance (Goose Island Anchorage):
51°56.25′ N, 128°25.00′ W (NAD 27)

Goose Island Anchorage Chart 3786;
anchor: 51°55.87′ N, 128°25.92′ W (NAD 27)

"Goose Point Cove" Chart 3921;
anchor: 51°04.92′ N, 127°30.58′ W (NAD 83)

Goose Spit Light Chart 3527;
position: (W. extremity of Goose Spit):
49°39.63′ N, 124°55.43′ W (NAD 27)

Gordon Islands Chart 3549;
safety position: 50°48.80′ N, 127°28.70′ W
(NAD 83)

Gordon River Chart 3647
entrance: 48°34.43′ N, 124°24.97′ W (NAD 27)

Gore Island Light Chart 3664;
position: (N. side of isl.):
49°39.30′ N, 126°23.48′ W (NAD 27)

Gore Island West Light Chart 3664;
position: (W. extremity of isl.):
49°38.95′ N, 126°25.93′ W (NAD 27)

Gorge Harbour Chart 3538, 3311;
public float: 50°05.95′ N, 125°01.19′ W
(NAD 83)

Gosse Bay Chart 3720;
anchor: 52°09.94′ N, 127°55.68′ W (NAD 27)

Gosse Passage Light Buoy U9
Chart 3476;
position: 48°42.02′ N, 123°24.24′ W
(NAD 27)

Gosse Point Light Chart 3495;
position: (off S. shore):
49°17.56′ N, 122°55.50′ W (NAD 83)

**Gossip Shoals Light and
Bell Buoy U47** Chart 3473;
position: (E. of shoal, off SE end of Gossip
Isl.): 48°53.10′ N, 123°18.28′ W (NAD 27)

Goudge Island Light Chart 3476;
position: (NW of Goudge Isl.):
48°41.33′ N, 123°23.68′ W (NAD 27)

Governor Rock Light Buoy U45
Chart 3442;
position: (E. of rock):
48°54.78′ N, 123°29.77′ W (NAD 27)

Gowgaia Point Light Chart 3864;
position: 52°24.02′ N, 131°35.28′ W
(NAD 27)

Gowlland Harbour Chart 3540;
anchor North Spit: 50°04.96′ N,
125°14.63′ W (NAD 83)

Gowlland Point Light Chart 3477;
position: (S. of Pender Island);
48°44.15′ N, 123°10.95′ W (NAD 27)

Grace Harbour Charts 3559, 3312, 3538;
anchor: 50°03.26′ N, 124°44.63′ W (NAD 27)

Grace Islands Light Chart 3526;
position: (SW extremity):
49°25.83′ N, 123°26.80′ W (NAD 83)

Grace Islet Light Chart 3478;
position: (SE extremity of isl.):
48°51.10′ N, 123°29.50′ W (NAD 83)

"Granby Cove" Chart 3920;
entrance: 55°22.85′ N, 129°49.05′ W;
anchor: 55°22.98′ N, 129°48.81′ W (NAD 83)

Granite Bay Chart 3539;
anchor: 50°14.37′ N, 125°18.37′ W (NAD 83)

"Granite Cove," Kiskosh Inlet
Chart 3743;
anchor: 53°33.84′ N, 129°20.76′ W (NAD 27)

Granite Falls Light Chart 3495;
position: (outer end of fill):
49°26.95' N, 122°51.81' W (NAD 83)

Granite Falls, Fairy Falls Chart 3495;
light: 49°26.96' N, 122°51.73' W (NAD 83)

Grant Anchorage
Charts 3710, 3728, 3737;
(uncharted) entrance:
52°28.83' N, 128°45.17' W (NAD 27)

Grant Bay Chart 3679;
anchor: 50°28.65' N, 128°05.83' W (NAD 83)

**Grant Reefs Bifurcation
Light Buoy QM** Chart 3311;
position: (S. extremity of reefs):
49°52.08' N, 124°45.97' W (NAD 83)

Grappler Inlet Chart 3646
entrance: 48°50.25' N, 125°08.00' W
(NAD 27)

Grappler Inlet Light Chart 3646;
position: (on pt, N. shore):
48°49.92' N, 125°07.55' W (NAD 83)

Grappler Rock Light Chart 3442;
position: (on rock):
48°56.35' N, 123°36.09' W (NAD 27)

Grappler Sound Chart 3547
east entrance: 50°55.15' N, 126°51.55' W
(NAD 83)

Grave Point Light Chart 3442;
position: (on the pt):
48°50.91' N, 123°35.48' W (NAD 27)

Grebe Cove Chart 3546;
anchor: 50°42.67' N, 126°38.20' W (NAD 83)

Grebe Islets Light Chart 3481;
position: 49°20.48' N, 123°16.48' W
(NAD 27)

Green Bay Chart 3730;
position: 52°20.53' N, 126°58.70' W (NAD 27)

Green Bay, Agamemnon Channel
Chart 3312;
anchor: 49°42.58' N, 124°04.78' W (NAD 27)

Green Cove Chart 3646 (inset);
position: 49°59.33' N, 125°58.86' W
(NAD 83)

Green Inlet Chart 3738;
entrance: 52°55.45' N, 128°29.90' W (NAD 27)

Green Island Anchorage Chart 3921;
entrance: 51°37.90' N, 127°50.35' W
anchor: 51°38.55' N, 127°50.40' W (NAD 83)

Green Island Chart 3959;
position (light): 54°34.10' N, 130°42.40' W
(NAD 83)

Green Island Light Chart 3959;
position: (SW side of isl.):
54°34.10' N, 130°42.40' W (NAD 83)

Greene Point Rapids Chart 3543;
Griffiths Islet light:
50°26.54' N, 125°30.33' W (NAD 27)

Greentop Islet Light Chart 3957;
position: (on islet):
54°10.65' N, 130°24.69' W (NAD 83)

Greenway Sound Chart 3547;
entrance: 50°51.70' N, 126°43.30' W;
marina float: 50°50.20' N, 126°46.40' W
(NAD 83)

Greenwood Point Light Chart 3686;
position: 50°30.59' N, 128°01.58' W
(NAD 83)

Grenville Channel Charts 3772, 3773;
south entrance: 53°22.00' N, 129°19.00' W;
north entrance: 53°55.25' N, 130°11.00' W
(NAD 27)

"Gribbell Islet Anchorage"
Chart 3955;
entrance: 54°20.00' N, 130°26.35' W;
anchor: 54°19.92' N, 130°26.42' W (NAD 27)

Grief Bay Chart 3934;
entrance: 51°24.90' N, 127°54.40' W;
anchor: 51°25.31' N, 127°54.68' W (NAD 83)

Grief Point East Light Chart 3311;
position: (end of S. breakwater Garden
Beach): 49°48.02' N, 124°31.07' W (NAD 83)

Grief Point Light Chart 3311;
position: (W. extremity of pt):
49°48.28' N, 124°31.50' W (NAD 83)

Griffin Passage (North Entrance)
Chart 3738;
entrance: 52°46.92' N, 128°20.82' W (NAD 27)

Griffin Passage (South Entrance)
Chart 3962;
south entrance: 52°35.14' N, 128°17.33' W;
south rapids (unsurveyed):
52°36.74' N, 128°17.60' W (NAD 27)

Griffin Point Light Chart 3739;
position: (on pt): 53°03.93' N, 128°32.90' W
(NAD 27)

Griffith Harbour Chart 3723 (inset);
anchor (Ford Rock):
53°35.98' N, 130°32.63' W (NAD 27)

Griffiths Islet Light Chart 3543;
position: (W. end of isl.):
50°26.54' N, 125°30.33' W (NAD 83)

Grimmer Bay (Port Washington)
Chart 3313, p. 8;
public float: 48°48.77' N, 123°19.23' W
(NAD 83)

Grindstone Point Light Chart 3955;
position: (on rock, N. of pt):
54°18.58' N, 130°23.14' W (NAD 27)

Growler Cove Chart 3546;
anchor: 50°32.45' N, 126°36.90' W (NAD 83)

Guise Bay Chart 3624;
anchor: 50°46.15' N, 128°24.50' W (NAD 27)

Gull Rocks Light Chart 3957;
position: (on highest rock of Gull Rocks
group): 54°07.96' N, 130°31.27' W (NAD 83)

Gunboat Bay Chart 3535 (inset);
anchor: (quarter mile east of Goat Islet)
49°37.65' N, 124°00.00' W;
(Oyster Bay) 49°37.80' N, 123°59.80' W
(NAD 27)

Gunboat Harbour Chart 3773;
entrance: 53°55.18' N, 130°08.51' W;
anchor: 53°55.38' N, 130°08.56' W (NAD 27)

"Gunboat Lagoon Cove"
Chart 3720 (inset);
entrance: 52°10.40' N, 127°58.75' W;
anchor: 52°10.59' N, 127°58.55' W (NAD 27)

Gunboat Passage Chart 3720 (inset);
east entrance: 52°09.64' N, 127°55.00' W;
west entrance:
52°10.20' N, 127°58.50' W (NAD 27)

"Gung Ho Bay," Banks Island
Chart 3742;
entrance: 53°10.79' N, 129°45.49' W;
anchor: 53°10.67' N, 129°46.55' W (NAD 27)

Gunner Inlet Chart 3649; inner basin
anchor: 49°10.08' N, 125°44.45' W (NAD 27)

Gurd Inlet Charts 3761, 3927;
entrance: 53°53.28' N, 130°39.16' W;
anchor: 53°53.18' N, 130°37.55' W (NAD 27)

**Gwaii Haanas/South Moresby
National Park Reserve** Chart 3807;
Dana Inlet entrance:
52°49.20' N, 131°38.80' W;
Logan Inlet entrance:
52°47.50' N, 131°38.90' W (NAD 27)

Gwayasdums Indian Reserve (Health Bay Indian Reserve) Chart 3546;
position: 50°41.70′ N, 126°36.10′ W
(NAD 83)

Gwent Cove Chart 3994;
entrance: 54°56.55′ N, 130°19.70′ W;
anchor: 54°56.57′ N, 130°20.00′ W (NAD 27)

Haaksvold Point Chart 3729;
position (light): 51°57.60′ N, 127°42.10′ W
(NAD 27)

Haaksvold Point Light Chart 3729;
position: (on pt): 51°57.97′ N, 127°42.60′ W
(NAD 27)

Haans Islet Light Chart 3894;
position: 53°02.32′ N, 131°41.21′ W
(NAD 27)

Haddington Island Light Chart 3546;
position: (N. side of isl.):
50°36.30′ N, 127°01.28′ W (NAD 83)

Haddington Island South Light
Chart 3546;
position: (S. side of isl.):
50°35.91′ N, 127°01.49′ W (NAD 83)

Haddington Passage Chart 3546;
light north side of island:
50°36.32′ N, 127°01.19′ W (NAD 83)

Haddington Reefs Light Buoy N20
Chart 3546;
position: (off reefs in Haddington Passage):
50°36.51′ N, 127°01.04′ W (NAD 83)

Haddington Reefs Pier Light
Chart 3546;
position: 50°36.46′ N, 127°00.63′ W
(NAD 83)

Hadley Bay Charts 3564 (inset), 3545;
anchor: 50°35.02′ N, 126°·12.40′ W (NAD 83)

Haggard Cove Chart 3668;
entrance: 48°57.75′ N, 125°01.43′ W (NAD 83)

"Hague Point Lagoon" Chart 3737;
entrance: 52°40.05′ N, 128°50.92′ W;
anchor: 52°40.19′ N, 128°51.87′ W (NAD 27)

Haig Rock Light Chart 3726;
position: (on rock off Tildesley Pt):
52°36.37′ N, 128°55.30′ W (NAD unknown)

Hakai Passage Chart 3784;
east entrance: 51°44.60′ N, 128°00.00′ W;
west entrance: 51°42.30′ N, 128°08.00′ W
(NAD 27)

"Half-Dome Waterfall" Chart 3552,
position: 5l°08.00′ N, 127°11.25′ W (NAD 27)

**Halibut Bank ODAS Light Buoy
46146** Chart 3463;
position: 49°20.40′ N, 123°43.60′ W
(NAD 27)

Halibut Bay Chart 3933;
entrance (east cove):
55°13.75′ N, 130°05.65′ W;
anchor (west shore):
55°13.20′ N, 130°05.85′ W (NAD 27)

Halkett Bay Chart 3526;
entrance: 49°26.75′ N, 123°19.65′ W (NAD 27)

"Hallet Island Cove" Chart 3891;
entrance: 53°12.95′ N, 132°14.30′ W;
anchor: 53°13.14′ N, 132°14.35′ W (NAD 83);

Hammond Bay (Pipers Lagoon)
Charts 3458, 3313;
anchor: 49°13.71′ N, 123°57.33′ W (NAD 27)

Hampden Bay Chart 3720;
anchor: 52°09.12′ N, 127°54.70′ W (NAD 27)

Hand Island Light Chart 3670;
position: 48°57.13′ N, 125°18.38′ W (NAD 83)

Hand Island Passage Light Chart 3670;
position: 48°56.69' N, 125°19.57' W (NAD 83)

Handfield Bay in Cameleon Harbour
Chart 3543;
anchor: 50°21.05' N, 125°18.91' W (NAD 27)

Hankin Cove Chart 3683;
anchor: 50°06.6' N, 127°13.6' W (NAD 27)

Hankin Rock Light Chart 3747;
position: (on rock, SW entrance to Beaver
Passage): 53°42.47' N, 130°24.60' W
(NAD 27)

Hanmer Island Light Chart 3717;
position: (S. end of isl., Arthur Passage):
54°03.35' N, 130°15.07' W (NAD 83)

**Hanmer Rocks Light and Whistle
Buoy D57** Chart 3957;
position: 54°18.90' N, 130°48.94' W
(NAD 83)

**Hanmer Rocks Light and
Whistle Buoy D62** Chart 3957;
position: 54°18.35' N, 130°48.54' W
(NAD 83)

Hanmer Rocks Light Chart 3957;
position: 54°19.45' N, 130°49.26' W
(NAD 83)

Hanna Channel Light Chart 3664;
position: (on pt, NE shore of Bligh Isl.):
49°40.59' N, 126°29.67' W (NAD 27)

"Hanna Point Bight" Chart 3624;
anchor: 50°40.13' N, 128°20.18' W (NAD 27)

Hansen Bay Chart 3624;
entrance: 50°44.04' N, 128°23.62' W (NAD 27)

Hanson Island Sector Light
Chart 3546;
position: (on Licka Pt):
50°33.96' N, 126°41.59' W (NAD 83)

Harbott Point Light Chart 3543;
position: (SW extremity of Stuart Isl.):
50°21.70' N, 125°08.07' W (NAD 83)

Harbour Rock Light Chart 3711;
position: (Stewart Narrows):
53°23.29' N, 129°16.54' W (NAD 27)

Harbourmaster Point Light
Chart 3785;
position: 52°03.90' N, 128°02.92' W
(NAD 27)

Hardy Bay Inner Light Buoy
Chart 3548;
position: 50°42.68' N, 127°29.23' W
(NAD 83)

Hardy Bay Inner Light Chart 3548;
position: (SE of wharf):
50°43.11' N, 127°28.98' W (NAD 83)

Hardy Bay Light Chart 3548;
position: (on rock, N. of wharf):
50°43.80' N, 127°28.98' W (NAD 83)

Hardy Inlet Chart 3932;
entrance: 51°41.40' N, 127°27.50' W
(NAD 83)

Harlequin Basin Chart 3785 (inset);
entrance: 51°52.40' N, 127° 52.90' W;
position: 51°53.25' N, 127°51.63' W
(NAD 27)

Harlequin Bay Chart 3549;
position: 50°50.57' N, 127°34.60' W
(NAD 83)

Harmony Islands Chart 3312;
anchor: 49°51.78' N, 124°00.80' W (NAD 27)

Harness Island Chart 3311;
anchor: 49°35.47' N, 124°01.03' W (NAD 83)

Haro Strait Cardinal Light Buoy VD
Chart 3440;
position: 48°27.10' N, 123°10.77' W
(NAD 27)

Harriet Harbour Chart 3809;
entrance: 52°18.40' N, 131°13.90' W;
anchor: 52°17.85' N, 131°13.30' W (NAD 27)

Harris Island Light Chart 3921;
position: (on island):
51°00.03' N, 127°33.75' W (NAD 83)

Harris Rock Light Chart 3909;
position: (on rock):
54°12.97' N, 130°45.89' W (NAD 83)

Hartley Bay Breakwater Light
Chart 3711;
position: 53°25.44' N, 129°14.96' W
(NAD 27)

Hartley Bay Charts 3711 (inset), 3743;
breakwater light: 53°25.44' N, 129°14.96' W
(NAD 27)

Harvell Islet Cove Chart 3921 (inset) for
vicinity of Nakwakto Rapids; Chart 3552;
anchor (north of Harvell Islet and the reefs):
51°06.19' N, 127°29.08' W (NAD 83)

Harvey Cove Charts 3686, 3679;
entrance: 50°25.76' N, 127°55.38' W
(NAD 83)

Harwood Bay Charts 3719 (inset), 3742;
entrance: 53°08.67' N, 129°33.96' W;
anchor (north basin):
53° 09.14' N, 129° 33.27' W;
anchor (east basin):
53° 08.82' N, 129° 33.11' W (NAD 27)

Harwood Island
Charts 3311, sheet 5, 3513;
Harwood Island south point:
49°50.20' N, 124°40.02' W (NAD 83)

Hastings Arm Chart 3920;
entrance (west fairway at Vadso Rocks):
55°22.87' N, 129°45.90' W;
entrance (east fairway at Brooke Shoal):
55°23.00' N, 129°43.50' W (NAD 83)

Haswell Bay Chart 3808;
entrance: 52°31.45' N, 131°36.15' W;
anchor: 52°30.53' N, 131°36.93' W (NAD 27)

Haswell Island Light Chart 3807;
position: (S. side of isl.):
52°51.65' N, 131°41.10' W (NAD 27)

Hattie Island Light Chart 3933;
position: (W. side of isl., Portland Canal):
55°17.25' N, 129°58.20' W (NAD 27)

**Havannah Channel (alternate
westbound route to Blackfish
Sound)** Chart 3564, 3545;
south entrance point:
50°31.70' N, 126°17.50' W (NAD 83)

Havannah Islets Light Chart 3564;
position: (southernmost rock of Havannah
Islets): 50°32.19' N, 126°15.04' W (NAD 83)

Havelock Rock Light Chart 3957;
position: (Chatham Sound, N. of Hunts Inlet):
54°05.91' N, 130°28.99' W (NAD 83)

Haven Cove Chart 3663;
anchor: 49°52.85' N, 126°46.66' W (NAD 27)

Hawk Bay Chart 3742;
entrance: 53°16.23' N, 129°22.30' W;
anchor (east): 53°16.27' N, 128°29.98' W
(NAD 27)

Hawkesbury Island Light Chart 3743;
position: (W. side of isl.):
53°37.12' N, 129°10.80' W (NAD 27)

"Hawkesbury Island North Bight"
Chart 3743;
anchor: 53°42.43' N, 128°53.49' W (NAD 27)

Hay Point Light Chart 3477;
position: (on pt, Bedwell Hbr):
48.44.70' N, 123°13.71' W (NAD 27)

Hayden Passage Chart 3648
NW entrance: 49°24.10' N, 126°07.55' W
SE entrance: 49°23.35' N, 126°05.65' W
(NAD 27)

Hayden Passage Light Chart 3643;
position: (on rock in middle of passage):
49°23.75' N, 126°06.30' W (NAD unknown)

Head Bay Chart 3664;
anchor: 49°47.59' N, 126°29.78' W (NAD 27)

Health Bay Chart 3546;
anchor: 50°40.94' N, 126°34.60' W (NAD 83)

Heard Island Chart 3549;
safety position: 50°48.88' N, 127°30.92' W
(NAD 83)

Heater Harbour Chart 3825;
entrance: 52°07.50' N, 130°59.85' W;
buoy: 52°07.50' N, 131°02.50' W (NAD 27)

Heathorn Bay Chart 3962; beach:
52°50.39' N, 128°08.02' W (NAD 27)

Hecate Bay Chart 3649;
entrance: 49°15' N, 125°56.5' W (NAD 27)

Hecate Channel Chart 3663;
entrance 49°53.50' N, 126°47.00' W (NAD 27)

Hecate Cove Chart 3681;
Kitten Islet anchor: 50°32.54' N, 127°35.52' W
(NAD 83)

Hecate Strait Charts 3853, 3902, 3802;
position (Bonilla Island Light):
53°29.55' N, 130°38.10' W;
position (White Rocks Light):
53°38.10' N, 130°38.80' W (NAD 27)

Hecate Strait ODAS Light Buoy
Chart 3902;
position: 53°37.00' N, 131°06.25' W
(NAD 27)

**Hecate Strait South ODAS
Light Buoy 46185** Chart 3726;
position: 52°25.20' N, 129°48.00' W
(NAD unknown)

Heelboom Bay Chart 3685;
anchor: 49°09.27' N, 125°47.73' W (NAD 83)

Helby Island Chart 3671;
48°51.09' N, 125°10.77' W (NAD 27)

Helen Bay Chart 3547;
anchor: 50°54.15' N, 126°56.17' W (NAD 83)

Helen Point Light Chart 3473;
position: (on pt): 48°51.47' N, 123°20.63' W
(NAD 27)

Helmcken Inlet Chart 3719 (inset);
entrance: 52°45.31' N, 129°04.52' W;
anchor (lagoon entrance):
52°45.88' N, 128°01.72' W (NAD 27)

Helmcken Island North Light
Chart 3544;
position: (N. side of isl.):
50°24.38' N, 125°52.49' W (NAD 83)

Helmcken Island Sector Light
Chart 3544;
position: (on drying rock):
50°23.93' N, 125°51.43' W (NAD 83)

Helmcken Island South Sector Light
Chart 3544;
position: (S. side of isl.):
50°23.64' N, 125°52.29' W (NAD 83)

Hemasila Inlet Chart 3934;
entrance: 51°31.30' N, 127°35.15' W
(NAD 83)

Hemming Bay Chart 3543;
anchor: 50°23.94′ N, 125°22.95′ W (NAD 27)

Henderson Bay Chart 3921:
position: 51°35.29′ N, 127°47.26′ W (NAD 83)

Henry Bay Chart 3527;
anchor: 49°36.13′ N, 124°50.00′ W (NAD 27)

"Herald Rock Cove" Chart 3548;
anchor: 50°43.94′ N, 127°24.76′ W (NAD 83)

Herbert Reefs Light Chart 3927;
position: (near northern end of reef,
Arthur Passage): 54°10.45′ N, 130°14.23′ W
(NAD 27)

Heriot Bay Charts 3538, 3539;
anchor: 50°06.30′ N, 125°12.80′ W (NAD 27)

Hernando Island
Charts 3311, sheet 5, 3538;
position Dog Bay:
49°59.65′ N, 124°54.00′ W (NAD 83)

"Heron Bay" Chart 3664;
anchor: 49°44.74′ N, 126°38.22′ W (NAD 27)

Herring Bay, Ruxton Island
Chart 3313, p. 18 (inset);
anchor: 49°05.00, 123°42.88′ W (NAD 83)

Hesquiat Harbour Chart 3640;
entrance buoy "ME":
49°23.12′ N, 126°25.80′ W (NAD unknown)

**Hesquiat Harbour Light and
Whistle Buoy ME** Chart 3640;
position: (at entrance to harbour):
49°23.12′ N, 126°25.80′ W (NAD unknown)

Hevenor Inlet Charts 3753 (inset), 3746;
entrance (outer): 53°37.60′ N, 130°04.00′ W;
entrance (lagoon):
53°38.96′ N, 129°55.33′ W;
anchor: 53°39.00′ N, 129°55.50′ W (NAD 27)

Hewitt Island Chart 3738 (inset);
anchor: 52°52.48′ N, 128°30.15′ W (NAD 27)

Hewitt Island Light Chart 3738;
position: 52°52.43′ N, 128°29.83′ W
(NAD 27)

Hewitt Rock Bifurcation Light Buoy
Chart 3738;
position: (off Hewitt Isl.):
52°52.14′ N, 128°29.24′ W (NAD 27)

Heydon Bay Chart 3543;
position: 50°34.83′ N, 125°34.00′ W
(NAD 27)

Hickey Cove Chart 3931;
position: 51°18.45′ N, 127°20.55′ W (NAD 83)

Hickey Point Light Chart 3545;
position: (on the pt):
50°26.98′ N, 126°05.12′ W (NAD 83)

Hidden Basin Chart 3311;
entrance: 49°41.95′ N, 124°11.70′ W
(NAD 83)

Hidden Inlet Chart 3994;
entrance: 54°56.80′ N, 130°19.75′ W;
anchor: 54°57.14′ N, 130°20.22′ W (NAD 27)

"Hideaway Bay" Chart 3743;
entrance: 53°48.70′ N, 128°53.40′ W;
anchor: 53°48.58′ N, 128°53.85′ W (NAD 27)

Hideaway Lodge Chart 3933; dock
position: 55°36.14′ N, 130°06.38′ W (NAD 27)

Hiekish Narrows Chart 3738 (inset);
south entrance: 52°48.75′ N, 128°26.35′ W;
north entrance: 52°53.10′ N, 128°30.10′ W;
Hewitt Rock light:
52°52.14′ N, 128°29.24′ W (NAD 27)

Hiekish Narrows Light Chart 3738;
position: (E. end of narrows):
52°49.63′ N, 128°26.75′ W (NAD 27)

Higgins Lagoon Charts 3737, 3728, 3726;
entrance: 52°29.80' N, 128°41.90' W (NAD 27)

Higgins Passage
Charts 3710 (inset), 3728, 3726, 3737;
east entrance: 52°27.60' N, 128°36.75' W;
anchor (east side of narrows):
52°28.88' N, 128°43.50' W;
anchor (west side of narrows):
52°28.50' N, 128°41.10' W;
west entrance (0.4 mile SW of Kipp Islet):
52°28.47' N, 128°47.27' W (NAD 27)

Hilton Point Light Chart 3743;
position: (on the pt):
53°48.93' N, 128°52.20' W (NAD 27)

Hippa Island Light Chart 3860;
position: (W. end of isl.):
53°32.70' N, 133°00.60' W (NAD 27)

"Hisnit Inlet Bay" Chart 3664;
anchor: 49°45.15' N, 126°30.73' W (NAD 27)

Hjorth Bay Chart 3539;
anchor: 50°10.63' N, 125°07.30' W (NAD 83)

Hkusam Bay Chart 3544;
position: 50°23.25' N, 125°55.40' W (NAD 83)

Hochstader Basin Chart 3787;
entrance: 52°06.35' N, 128°16.90' W
(NAD 27)

"Hochstader Cove" Chart 3787;
entrance: 52°06.96' N, 128°15.80' W;
anchor: 52°06.79' N, 128°15.68' W (NAD 27)

Hocking Point Light Chart 3668;
position: (on extremity of pt, W. side of
Alberni Inlet): 49°05.28' N, 124°49.88' W
(NAD 83)

Hodgson Cove Chart 3721;
entrance (outer): 53°27.00' N, 129°52.85' W;
entrance (inner): 53°27.40' N, 129°52.52' W;
anchor: 53°27.48' N, 129°51.94' W (NAD 27)

Hodgson Reefs Chart 3959;
south end of reef: 54°22.35' N, 130°31.73' W
(NAD 83)

**Hodgson Reefs Light and
Whistle Buoy D84** Chart 3957;
position: (W. of reefs):
54°23.07' N, 130°32.48' W (NAD 83)

Hohm Island Light Chart 3668;
position: (on isl.): 49°13.66' N, 124°49.54' W
(NAD 83)

Holberg Inlet Chart 3679
entrance: 50°33.90' N, 127°33.60' W
(NAD 83)

Hole in the Wall Chart 3537;
west entrance: 50°17.85' N, 125°12.75' W;
east entrance: 50°19.80' N, 125°07.60' W
(NAD 27)

Hole in the Wall West Light
Chart 3537;
position: (on shore):
50°17.98' N, 125°12.42' W (NAD 27)

Holiday Island Light Chart 3992;
position: (N. extremity of isl.):
54°37.38' N, 130°45.50' W (NAD unknown)

Holland Rock Light Chart 3957;
position: (on southern extremity of rock):
54°10.34' N, 130°21.66' W (NAD 83)

Holmes Inlet Chart 3648;
anchor: 49°26' N, 126°14' W (NAD 27)

"Holmes Inlet Nook" Chart 3648;
anchor: 49°27.72' N, 126°14.06' W (NAD 27)

"Holmes Point Cove" Chart 3552;
anchor: 51°05.66' N, 127°27.69' W (NAD 27)

Home Bay Chart 3742;
entrance: 53°16.60 ' N, 129°05.40' W
(NAD 27)

Home Bay Chart 3934;
entrance: 51°23.88' N, 127°42.48' W (NAD 83)

Home Bay Charts 3740, 3742;
anchor: 53°16.29' N, 129°04.77' W (NAD 27)

Hood Point Chart 3526;
Finisterre light: 49°25.10' N, 123°18.43' W
(NAD 27)

Hoop Bay Chart 3550;
position: 51°13.30' N, 127°46.68' W
(NAD 83)

"Hoot-la-Kootla" Chart 3648;
anchor south: 49°21.69' N, 126°13.80' W;
anchor north: 49°21.93' N, 126°13.72' W
(NAD 27)

Hope Point Light Chart 3526;
position: (S. end Gambier Isl.):
49°25.85' N, 123°22.00' W (NAD 83)

**Horda Shoals Bifurcation Light
Buoy UD** Chart 3478;
position: (SE of shoals):
48°49.87' N, 123°24.95' W (NAD 83)

Horn Island Light Chart 3859;
position: (N. tip of isl.):
52°46.15' N, 132°03.38' W (NAD 27)

Hornby Island Ferry Landing Light
Chart 3527;
position: (SE of Shingle Spit):
49°30.68' N, 124°42.20' W (NAD 27)

Hornby Point Light Chart 3855;
position: 52°09.29' N, 131°06.33' W
(NAD 27)

Horsefly Cove Chart 3738;
anchor: 52°55.28' N, 128°28.95' W (NAD 27)

Horswell Channel Chart 3458;
Clark Rock Light: 49°13.53' N, 123°56.42' W
(NAD 27)

**Horswell Rock East Cardinal Light
Buoy PL** Chart 3457;
position: (off reef, off Horswell Bluff):
49°12.73' N, 123°55.88' W (NAD 27)

Horton Bay Chart 3313, p. 12;
anchor: 48°49.75' N, 123°15.00' W (NAD 83)

Hoskyn Channel Chart 3539;
50°07.50' N, 125°09.50' W (NAD 83)

Hospital Bay Chart 3535 (inset);
public floats: 49°37.94' N, 124°01.85' W
(NAD 27)

Hospital Rock Light Buoy U30
Chart 3475;
position: (E. of the rock):
48°55.78' N, 123°42.70' W (NAD 27)

Hot Springs Cove Charts 3643, 3648;
anchor: 49°21.93' N, 126°15.95' W
(NAD unknown)

Hotspring Island Chart 3808;
position (springs): 52°34.45' N, 131°26.42' W;
anchor (east side): 52°34.64' N, 131°25.90' W
(NAD 27)

Houston River Light Chart 3664;
position: (NE of river entrance):
49°38.50' N, 126°17.48' W (NAD 27)

Houston Stewart Channel Chart 3855;
east entrance: 52°09.55' N, 131°02.30' W;
west entrance: 52°06.65' N, 131°09.50' W
(NAD 27)

Hovel Bay Charts 3542, 3312;
position: 50°39.80' N, 124°51.50' W (NAD 83)

Howe Sound Chart 3526;
Point Atkinson light:
49°19.83' N, 123°15.80' W (NAD 27)

Hoy Bay (Hopetown Passage)
Chart 3547;
enter: 50°55.33' N, 126°50.18' W (NAD 83)

"Hudson Bay Passage Cove"
Chart 3959;
entrance: 54°30.96' N, 130°50.48' W;
anchor: 54°30.90' N, 130°50.94' W (NAD 83)

Hudson Island North Light Chart 3477;
position: 48°57.97' N, 123°40.40' W
(NAD 27)

Hudson Rocks Light Chart 3458;
position: (on summit of SW islet):
49°13.43' N, 123°55.59' W (NAD 83)

Humpback Bay Chart 3544;
anchor: 50°21.67' N, 125°41.48' W (NAD 83)

Humpback Bay, Porcher Island
Chart 3956;
entrance: 54°05.50' N, 130°23.20' W;
anchor: 54°05.26' N, 130°23.99' W (NAD 83)

Hunt Inlet Charts 3909 (inset), 3956;
entrance: 54°05.16' N, 130°27.25' W;
public float: 54°04.12' N, 130°26.68' W;
anchor: 54°02.78' N, 130°26.32' W (NAD 83)

Hunt Rock Light Buoy N35 Chart 3549;
position: 50°54.27' N, 127°40.55' W
(NAD 83)

Hunter Channel Charts 3787, 3786;
north entrance: 52°04.80' N, 128°07.30' W;
south entrance: 51°59.80' N, 128°11.00' W
(NAD 27)

Hunter Channel Complex Chart 3787;
anchor: 52°03.06' N, 128°07.93' W (NAD 27)

"Hurricane Anchorage" Chart 3784;
anchor: 51°50.21' N, 128°12.62' W (NAD 27)

Hutton Inlet Chart 3808;
entrance: 52°31.50' N, 131°31.90' W;
anchor: 52°30.00' N, 131°34.08' W (NAD 27)

"Hutton Island Cove" Chart 3808;
anchor: 52°30.88' N, 131°31.43' W (NAD 27)

Hyacinthe Bay Charts 3539, 3312;
entrance: 50°07.10' N, 125°13.00' W
(NAD 83)

Hyde Rock Light Chart 3920;
position: (on rock, E. side of Liddle Channel):
55°23.72' N, 129°41.07' W (NAD 83)

Hyder, Alaska Chart 3794; float
position: 55°54.24' N, 130°00.56' W (NAD 27)

Hyndman Reefs Light Chart 3720;
position: (on reef): 52°15.75 N,
128°14.55' W (NAD 27)

Iceberg Bay Chart 3920;
entrance: 54°57.50' N, 129°56.50' W
(NAD 83)

Idol Point Light Chart 3787;
position: (extremity of pt):
52°14.32' N, 128°16.52' W (NAD 27)

Ikeda Cove Chart 3809;
entrance: 52°18.80' N, 131°08.00' W;
anchor: 52°17.80' N, 131°09.30' W (NAD 27)

Ildstad Island Chart 3679;
anchor: 50°31.10' N, 127°42.00' W (NAD 83)

Ildstad Islands Light Chart 3679;
position: (SE extremity of E. Ildstad Isl.):
50°30.94' N, 127°41.85' W (NAD 83)

Illahie Inlet Chart 3921;
entrance: 51°37.94' N, 127°49.76' W
(NAD 83)

Imperieuse Rock Light Buoy P21
Chart 3459;
position: (N. of rock):
49°15.73' N, 124°07.38' W (NAD 27)

Indian Arm Chart 3495;
entrance: 49°18.00' N, 122°56.50' W
(NAD 83)

Indian Arm Marine Park Chart 3495;
Racoon Island anchor:
49°20.52′ N, 122°54.26′ W (NAD 83)

Indian Island Charts 3934, 3931;
anchor: 51°17.15′ N, 127°38.38′ W (NAD 83)

Indian Reserve Light Chart 3646;
position: (N. of Hyphocus Isl.):
48°56.09′ N, 125°31.65′ W (NAD 83)

"Ingram Bay" Chart 3940;
entrance: 52°37.80′ N, 128°02.20′ W;
anchor: 52°37.63′ N, 128°01.72′ W (NAD 83)

Inner (Easterly) Light Chart 3491;
position: (S. side of channel at turn in North
Arm Jetty): 49°13.66′ N, 123°13.62′ W
(NAD 83)

Inner Basin Chart 3663;
entrance: 49°48.0′ N, 126°49.5′ W (NAD 27)

"Inner Warrior Cove" Chart 3784;
entrance: 51°49.90′ N, 127°53.50′ W,
anchor: 51°50.45′ N, 127°52.49′ W (NAD 27)

**International Boundary
Range A Light** Chart 3499;
position: (on shore northerly of Pt. Roberts):
49°00.13′ N, 123°05.33′ W (NAD 27)

**International Boundary
Range B Light** Chart 3463;
position: (on shore):
49°00.13′ N, 123°02.03′ W (NAD 27)

**International Boundary
Range C Light** Chart 3463;
position: (on shore):
49°00.15′ N, 122°46.83′ W (NAD 27)

Inverness Passage Chart 3717;
west entrance (0.95 mile SE buoy "D18"):
54°09.40′ N, 130°17.65′ W; (course 356° M
2.4 miles to Mathews Rock Buoy) (NAD 27)

Inverness Passage Light Chart 3717;
position: (mouth of Skeena River):
54°11.95′ N, 130°15.71′ W (NAD 83)

**Iphigenia Point (Langara Island)
Light** Chart 3895;
position: (on pt, SW extremity of Langara
Isl.): 54°11.42′ N, 133°00.65′ W (NAD 27)

Ire Inlet Chart 3746;
entrance: 53°30.10′ N, 129°54.45′ W;
anchor: 53°29.99′ N, 129°56.32′ W (NAD 27)

Irish Bay Chart 3313, p. 12;
anchor: 48°49.05′ N, 123°12.42′ W (NAD 83)

Iron Mine Bay Chart 3641;
anchor: 48°20.31′ N, 123°42.22′ W (NAD 27)

"Iroquois Cove" Chart 3787;
anchor: 52°02.53′ N, 128°19.18′ W (NAD 27)

**Iroquois Passage and Colburne
Passage** Chart 3313, pp. 7, 6; Chart 3476;
Goudge Island (northwest) light:
48°41.33′ N, 123°23.68′ W;
Colburne Passage south light:
48°41.85′ N, 123°25.47′ W (NAD 83)

Irving Cove Chart 3649;
anchor: 49°11.47′ N, 125°37.66′ W (NAD 27)

Isabel Bay Charts 3559, 3312, 3538;
anchor: 50°03.33′ N, 124°43.71′ W (NAD 27)

Isabella Island Light Chart 3476;
position: (SE extremity of isl.):
48°43.73′ N, 123°25.75′ W (NAD 27)

**"Island (35) Cove,"
Bullock Channel** Chart 3940;
anchor: 52°24.94′ N, 128°04.99′ W (NAD 83)

**"Island (45) Cove,"
Bullock Channel** Chart 3940;
anchor: 52°25.48′ N, 128°05.53′ W (NAD 83)

Island Cove Chart 3649;
anchor: 49°08.80′ N, 125 45.80′ W (NAD 27)

Island Harbour Chart 3670
harbour entrance: 48°54.42′ N,
125°15.45′ W (NAD 27)

Ivory Island Chart 3710 (inset);
light: 52°16.17′ N, 128°24.30′ W;
anchor: 52°16.67′ N, 128°23.49′ W (NAD 27)

Ivory Island Light Chart 3710;
position: (Robb Pt, Milbanke Sound):
52°16.18′ N, 128°24.30′ W (NAD 27)

Jacinto Islands Light Chart 3723;
position: (SE end of isl.):
52°56.50′ N, 129°36.74′ W (NAD 27)

Jacinto Point Light Chart 3959;
position: (S. pt of Zayas Isl.):
54°34.78′ N, 131°04.50′ W (NAD 83)

Jack Point Light Chart 3457;
position: 49°10.05′ N, 123°53.53′ W (NAD 27)

Jackscrew Island Light Chart 3442;
position: 48°57.05′ N, 123°35.10′ W
(NAD 27)

Jackson Narrows
Charts 3711 (inset), 3734 (inset);
east narrows entrance:
52°31.33′ N, 128°17.60′ W;
west narrows entrance:
52°31.47′ N, 128°18.25′ W (NAD 27)

Jackson Passage to Klemtu
Charts 3711, 3734;
narrows east entrance:
52°31.35′ N, 128°16.80′ W;
west entrance: 52°32.75′ N, 128°26.50′ W
(NAD 27)

Jackson Rock Light Chart 3470;
position: (W. of rock):
48°45.33′ N, 123°25.97′ W (NAD unknown)

Jacobsen Bay Charts 3730, 3729;
position: 52°18.88′ N, 127°09.23′ W (NAD 27)

James Bay Chart 3313, p. 10;
anchor: 48°50.47′ N, 123°23.85′ W (NAD 83)

James Bay Chart 3962;
entrance: 52°41.45′ N, 128°12.15′ W;
anchor: 52°42.68′ N, 128°12.65′ W (NAD 27)

James Island Light Chart 3441;
position: (off NW pt of isl.):
48°37.06′ N, 123°22.71′ W (NAD 27)

James Point Light Chart 3547;
position: (on pt, W. side of entrance to Wells
Pass): 50°50.32′ N, 126°58.61′ W (NAD 83)

Jane Cove Chart 3785;
entrance: 52°03.90′ N, 128°04.40′ W;
anchor: 52°03.11′ N, 128°03.52′ W (NAD 27)

Jaques/Jarvis Lagoon Chart 3670;
anchor: 48°55.35′ N, 125°16.49′ W (NAD 83)

"Jarvis West Cove" Chart 3670
position: 48°55.57′ N, 125°17.34′ W (NAD 27)

Jeannette Islands Charts 3548, 3574;
south entrance: 50°55.32′ N, 127°24.04′ W;
anchor: 50°55.72′ N, 127°24.24′ W (NAD 83)

Jeannette Islands Light Chart 3550;
position: (on largest isl.):
50°55.29′ N, 127°24.76′ W (NAD 83)

Jedediah Bay *Chart* 3312, 3512;
anchor 49°30.19′ N, 124°12.73′ W (NAD 27)

Jedway Bay Chart 3809;
entrance: 52°18.10′ N, 131°15.80′ W;
anchor: 52°17.25′ N, 131°15.43′ W (NAD 27)

**Jeffrey Rock Light and
Whistle Buoy E64** Chart 3726;
position: (SW of rock):
52°27.28′ N, 128°49.40′ W (NAD unknown)

Jennis Bay Chart 3547;
anchor: 50°54.88' N, 127°02.06' W (NAD 83)

Jenny Inlet Chart 3781;
entrance: 52°16.00' N, 127°37.50' W;
anchor: 52°13.75' N, 127°35.60' W (NAD 27)

Jervis Inlet Chart 3514
entrance: 49°44.00' N, 124°14.60' W (NAD 27)

Jesse Falls Chart 3743;
position: 53°50.00' N, 128°51.50, W (NAD 27)

Jesse Island Light Chart 3457;
position: (eastern extremity of isl.):
49°12.48' N, 123°56.51. W (NAD 27)

Jesus Pocket Chart 3552;
anchor: 51°05.03' N, 126°53.77' W (NAD 27)

Jeune Landing Charts 3681, 3679;
entrance: 50°26.45' N, 127°29.90' W
(NAD 83)

Jewitt Cove Chart 3664;
anchor 49°41.43' N, 126°36.07' W (NAD 27)

Jewitt Cove Light Chart 3664;
position: 49°41.95' N, 126°36.00' W
(NAD 27)

Joan Point Light Chart 3475;
position: (on pt, Dodd Narrows):
49°08.15' N, 123°49.03' W (NAD 27)

Joassa Channel Chart 3787;
south entrance: 52°07.20' N, 128°18.70' W
(NAD 27)

Joe Cove Charts 3547, 3515;
anchor: 50°44.85' N, 126°39.43' W (NAD 83)

Joe's Bay Chart 3921;
anchor: 51°38.97' N, 127°45.54' W (NAD 83)

"Joes Bay" Chart 3670;
anchor: 48°54.93' N, 125°19.43' W (NAD 83)

Johnson Channel Chart 3720;
south entrance: 52°11.70' N, 127°53.20' W;
north entrance: 52°18.20' N, 127°56.50' W
(NAD 27)

Johnson Lagoon Chart 3683;
entrance: 50°10' N, 127°39' W (NAD 27)

Johnston Bay Chart 3932;
position: 51°29.50' N, 127°32.70' W (NAD 83)

Johnstone Strait
Charts 3543, 3544, 3545, 3546;
Chatham Point light:
50°20.02' N, 125°26.34' W (NAD 27);
Blinkhorn light: 50°32.63', 126°46.85'
(NAD 83)

Jones Cove Chart 3934;
entrance: 51°15.10' N, 127°45.90' W;
anchor: 51°14.88' N, 127°45.90' W (NAD 83)

Jorgensen Harbour Chart 3734;
entrance: 52°38.50' N, 128°35.05' W;
anchor: 52° 38.66' N, 128° 34.94' W (NAD 27)

Jorkins Point Light Chart 3734;
position: 52°26.43' N, 128°29.08' W
(NAD 27)

Juan Perez Sound Chart 3808; north
entrance: 52°34.30' N, 131°36.30' W
(NAD 27)

Juggins Bay Chart 3920;
position: 55°21.92' N, 129°47.17' W (NAD 83)

Julia Passage Chart 3670, 3668;
anchor: 48°57.52' N, 125°12.77' W (NAD 83)

Julian Cove Chart 3679;
anchor: 50°29.08' N, 127°36.48' W (NAD 83)

Junction Point Light Chart 3555;
position: (E. side of Cortes Isl.):
50°08.39' N, 124°53.60' W (NAD 27)

Kagan Bay Chart 3891;
entrance: 53°12.85' N, 132°08.60' W
(NAD 83)

Kains Island Quatsino Light
Chart 3686;
light: 50°26.48' N, 128°01.85' W (NAD 83)

Kakushdish Harbour, "Gullchuck"
Chart 3785;
entrance: 52°09.50' N, 128°02.70' W;
anchor: 52°08.97' N, 128°00.85' W (NAD 27)

Kamils Anchorage Charts 3682, 3683:
anchor: 50°00.2' N, 127°23.6' W (NAD 27)

Kanaka Bluff Light Chart 3476;
position: (W. pt of Portland Isl.):
48°43.57' N, 123°23.14' W (NAD 27)

Kanish Bay Chart 3539;
entrance: 50°15.50' N, 125°22.50' W (NAD 83)

"Kanish Bay Cove" Chart 3539;
anchor: 50°14.93' N, 125°18.60' W (NAD 83)

Karouk Island Light Chart 3682;
position: 50°04.63' N, 127°09.57' W
(NAD 27)

Kashutl Inlet Chart 3682
entrance: 50°06.40' N, 127°15.00' W
(NAD 27)

"Kayak Cove" Chart 3648;
entrance: 49°21.3' N, 126.13.7' W (NAD 27)

"Kayak Cove" Chart 3786;
anchor: 51°53.91' N, 128°13.88' W (NAD 27)

Keith Anchorage Chart 3784;
entrance: 51°39.05' N, 128°05.25' W (NAD 27)

Kelkpa Point Light Chart 3729;
position: (on pt): 52°07.28' N, 127°36.28' W
(NAD 27)

Kelp Passage Cove Chart 3927;
anchor: 54°00.88' N, 130°15.69' W (NAD 27)

Kelp Reef Light Buoy D63 Chart 3955;
position: 54°19.91' N, 130°27.67' W
(NAD 27)

Kelp Reefs Light Chart 3440;
position: (on the NE, reef, Haro Strait):
48°32.87' N, 123°14.13' W (NAD 27)

Kelpie Point Light Chart 3784;
position: (on the pt):
51°44.02' N, 127°59.70' W (NAD 27)

Kelsey Bay Chart 3544;
public float: 50°23.80' N, 125°57.57' W
(NAD 83)

Kemano Bay Charts 3736 (inset), 3745;
entrance: 53°28.30' N, 128°08.30' W;
yacht club entrance:
53°28.79' N, 128°07.51' W;
anchor (southeast entrance Bluff):
53°27.90' N, 128°07.15' W (NAD 83)

Kendrick Inlet Light Chart 3665;
position: (on rock):
49°43.12' N, 126°38.40' W (NAD 27)

Kennedy Cove ("Brewster's Bay")
Chart 3649;
anchor: 49°08.74' N, 125°40.10' W (NAD 27)

Kennedy Island Light Chart 3927;
position: (W. side of isl.):
54°00.40' N, 130°12.40' W (NAD 27)

Kenneth Passage Chart 3547
NW entrance: 50°57.03' N, 126°48.90' W
SE entrance: 50°55.96' N, 126°46.47' W
(NAD 83)

Kent Inlet Chart 3719;
entrance: 52°42.36' N, 129°00.74' W;
anchor (inner basin):
52°43.24' N, 128°58.78' W (NAD 27)

Kersey Point Light Chart 3743;
position: (on pt): 53°46.58' N, 128°51.55' W
(NAD 27)

Keswar Inlet Chart 3747;
entrance; 53°38.43' N, 130°21.76' W;
anchor (inner basin):
53°38.43' N, 130°20.48' W (NAD 27)

Keswar Point Light Chart 3747;
position: (S. of Keswar Pt):
53°37.58' N, 130°20.97' W (NAD 27)

Kettle Inlet Charts 3737, 3724;
entrance: 52°42.80' N, 129°17.30' W;
anchor (east shore):
52°41.88' N, 129°13.91' W (NAD 27)

Keyarka Cove Chart 3747;
entrance: 53°36.25' N, 130°21.40' W (NAD 27)

Khutze Inlet Chart 3739;
entrance: 53°04.90; N, 128°32.80' W;
anchor (east side of spit):
53°05.42' N, 128°30.95' W (NAD 27)

**Khutzeymateen
Inlet Head** Chart 3994;
anchor: 54°36.30' N, 129°55.90' W (NAD 27)

Khutzeymateen Inlet Chart 3994;
entrance: 54°43.20' N, 130°13.30' W (NAD 27)

Kid Bay Chart 3738;
entrance: 52°48.00' N, 128°23.70' W (NAD 27)

Kiewnuggit Light Chart 3772;
position: (on northwestern rock of Morning
Reef, Grenville Ch.):
53°40.72' N, 129°46.00' W (NAD 27)

Kilbella Bay Chart 3932;
position: 51°42.20' N, 127°20.40' W (NAD 83)

Kildala Arm Chart 3743;
entrance 53° 51.50' N, 128° 40.00' W
(NAD 27)

Kildidt Inlet and Lagoon Chart 3786;
inlet south entrance:
51°52.32' N, 128°06.78' W (NAD 27)

Kildidt Sound Chart 3784;
southwest entrance:
51°47.00' N, 128°12.00' W (NAD 27)

Killam Bay Chart 3312;
breakwater: 49°47.07' N, 123°55.21' W
(NAD 27)

Kiln Bay, Chapple Inlet Charts 3719
(inset), 3737;
anchor: 52°57.95' N, 129°08.83' W (NAD 27)

Kiltik Cove Chart 3785;
entrance: 51°53.80' N, 128°00.00' W;
anchor (south cove):
51°53.71' N, 128°00.63' W (NAD 27)

Kiltuish Inlet Chart 3743;
entrance: 53°24.40' N, 128°31.70' W;
anchor (outside
narrows): 53°23.82' N, 128°30.23' W
(NAD 27)

Kimsquit Bay Chart 3730;
position: 52°50.20' N, 126°58.00' W (NAD 27)

Kincolith Chart 3920;
boat harbour entrance:
54°59.83' N, 129°58.66' W (NAD 83)

Kindakun Point Light Chart 3869;
position: 53°18.92' N, 132°46.20' W (NAD 27)

"King Bay" Chart 3552;
position: 51°06.71' N, 127°31.02' W (NAD 27)

Kingcome Inlet Chart 3515;
entrance: 50°52.30' N, 126°37.20' W
(NAD 83)

Kingcome Point Light Chart 3740;
position: (N. extremity of pt):
53°17.95' N, 128°54.38' W (NAD 27)

Kingkown Inlet Chart 3753 (inset);
entrance (Reverie Passage):
53°28.00' N, 130°26.00' W;
position (Byers Bay):
53°31.20' N, 130°22.50' W (NAD 27)

Kingsley Point Coves Chart 3787;
entrance: 52°07.60' N, 128°18.35' W;
anchor (south cove):
52°07.64' N, 128°18.68' W;
anchor (north cove):
52°07.78' N, 128°18.57' W (NAD 27)

Kingui Island Light Chart 3894;
position: (SW extremity of isl.):
53°01.50' N, 131°37.85' W (NAD 27)

Kinsman Inlet Chart 3786;
entrance: 51°55.88' N, 128°11.33' W (NAD 27)

"Kipling Cove," Harmony Islands
Chart 3312;
anchor: 49°51.96 N, 124°01.00' W (NAD 27)

Kirkland Island Breakwater Light
Chart 3490;
position: 49°06.90' N, 123°05.08' W
(NAD 83)

Kirkland Island Range Light
Chart 3490;
position: 49°06.81' N, 123°05.15' W
(NAD 83)

Kirkland Island West Light Chart 3490;
position: (N. side of isl.):
49°06.72' N, 123°06.05' W (NAD 83)

Kisameet Bay Chart 3785;
entrance: 51°56.60' N, 127°54.30' W,
anchor: 51°58.10' N, 127°52.80' W (NAD 27)

Kiskosh Inlet Chart 3743;
entrance: 53°30.90' N, 129°13.90' W;
anchor (on bar): 53°30.86' N, 129°15.80' W
(NAD 27)

Kitamaat Village Chart 3736 (inset);
position (floating breakwater):
53°58.06' N, 128°39.12' W (NAD 83)

Kitasu Bay Charts 3737, 3726;
entrance: 52°33.50' N, 128°47.00' W
(NAD 27)

Kitimat Arm Chart 3743;
entrance: 53°49.00' N, 128°50.00' W (NAD 27)

Kitimat Harbour Chart 3736 (inset);
yacht club entrance:
53°59.96' N, 128°41.58' W (NAD 83)

Kitkatla Channel
Charts 3747, 3761, 3927;
south entrance: 53°47.10' N, 130°20.80' W;
north entrance: 53°51.50' N, 130°32.75' W
(NAD 27)

Kitkatla Charts 3761, 3747, 3921;
public floats: 53°47.74' N, 130°26.18' W
(NAD 27)

Kitkatla Inlet Charts 3761, 3927;
entrance: 53°51.50' N, 130°32.75' W (NAD 27)

Kitkiata Inlet Chart 3743;
entrance: 53°37.40' N, 129°14.00' W (NAD 27)

Kitlope Bight Chart 3745;
anchor: 53°15.55' N, 127°54.22' W (NAD 27)

Kitlope Valley Chart 3745; drying flat
position: 53°15.00' N, 128°54.00' W (NAD 27)

Kitsault Chart 3920;
position: 55°27.57' N, 129°28.88' W (NAD 83)

Kitsaway Anchorage Chart 3743;
entrance: 53°37.80' N, 128°52.10' W;
anchor: 53°36.45' N, 128°52.51' W (NAD 27)

Kitsilano Light Buoy Q52 Chart 3493;
position: 49°16.90' N, 123°08.90' W
(NAD 83)

"Kittyhawk Cove" Chart 3784;
anchor: 51°49.80' N, 128°11.20' W (NAD 27)

Kiwash Cove Chart 3784;
entrance: 51°50.70' N, 127°53.30' W;
anchor (northwest corner):
51°50.98' N, 127°52.91' W (NAD 27)

Kiwash Island Light Chart 3797;
position: (S. end of isl.):
51°51.66' N, 127°53.53' W (NAD unknown)

Klaquaek Channel, "The Lake"
Chart 3934;
southwest entrance:
51°27.25' N, 127°44.00' W;
northeast entrance:
51°30.60' N, 127°40.82' W (NAD 83)

Klaskino Anchorage Chart 3651;
anchor west of buoys:
50°18.18' N, 127°49.00' W,
anchor north nook:
50°18.36' N, 127°48.95' W (NAD 83)

Klaskino Inlet Charts 3680, 3651;
Scouler Entrance way-point:
50°18.10' N, 127°50.00' W (NAD 27)

Klaskish Anchorage Chart 3680;
anchor: 50°13.90' N, 127°45.82' W (NAD 27)

Klaskish Basin Chart 3680;
anchor: 50°15.37' N, 127°43.85' W (NAD 27)

Klaskish Inlet Chart 3680;
McDougal Island way-point:
50°14.26' N, 127°47.52' W (NAD 27)

**Klekane Inlet, Marmot Cove
and Scow Bay** Chart 3739;
entrance: 53°10.95' N, 128°38.80' W (NAD 27)

Klemtu Anchorage, Clothes Bay
Charts 3711 (inset), 3734;
anchor (Clothes Bay):
52°34.42' N, 128°30.92' W (NAD 27)

Klemtu Passage
Charts 3711 (inset), 3734;
south entrance: 52°33.15' N, 128°29.60' W;
north entrance: 52°36.65' N, 128°31.25' W
(NAD 27)

Klewnuggit Inlet Chart 3772;
entrance: 53°41.40' N, 129°45.20' W (NAD 27)

Kliktsoatli Harbour Chart 3785 (inset);
entrance: 52°09.70' N, 128°05.60' W;
anchor: 52°08.67' N, 128°04.48' W (NAD 27)

Kloiya Bay Chart 3955;
position: 54°15.06' N, 130°11.78' W
(NAD 27)

Knapp Island Light Buoy U10
Chart 3476;
position: (W. of isl.):
48°41.95' N, 123°24.20' W (NAD 27)

Knight Inlet Charts 3545, 3546, 3515
west entrance: 50°38.00' N, 126°44.20' W
(NAD 83)

Knox Bay Chart 3543;
anchor: 50°23.47' N, 125°37.14' W (NAD 27)

Koeye River Chart 3784;
entrance: 51°46.48' N, 127°52.88' W
anchor: 51°46.42' N, 127°52.58' W (NAD 27)

Kokwina Cove Chart 3681;
anchor: 50°31.45' N, 127°34.40' W (NAD 83)

Kooh Rock Light Buoy Y51 Chart 3671;
position: (off rock):
48°54.23' N, 125°04.03' W (NAD 27)

Kooryet Bay Chart 3741;
entrance: 53°20.35' N, 129°51.35' W;
anchor: 53°20.13' N, 129°51.82' W (NAD 27)

Koprino Harbour Chart 3679
entrance: 50°29.40' N, 127°51.70' W (NAD 83)

"Koprino River (West Side)"
Chart 3679;
anchor: 50°30.31' N, 127°51.55' W (NAD 83)

Koskimo Island Light Chart 3679;
position: 50°28.63' N, 127°51.29' W (NAD 83)

Koskimo Islands Chart 3679;
anchor: 50°28.17' N, 127°51.25' W (NAD 83)

Kostan Inlet Chart 3808;
entrance: 52°35.30' N, 131°40.50' W;
anchor: 52°34.82' N, 131°42.25' W (NAD 27)

"Kostan Point Cove" Chart 3808;
entrance: 52°35.00' N, 131°40.35' W;
anchor: 52°34.94' N, 131°40.10' W (NAD 27)

Kshwan River Chart 3933;
position: 55°37.40' N, 129°48.48' W (NAD 27)

**Kuhushan Point (Salmon
Point Marina)** Chart 3513;
Kuhushan Point light:
49°53.33' N, 125°07.30' W (NAD 27)

Kuhushan Point Light Chart 3513;
position: (on the pt):
49°53.33' N, 125°07.30' W (NAD 27)

Kulleet Bay Chart 3443 (or 3313, p. 18);
anchor: 49°01.23' N, 123°47.03' W (NAD 27)

Kultus Cove Chart 3679;
anchor: 50°29.00' N, 127°37.00' W (NAD 83)

Kumealon Inlet Chart 3773;
entrance: 53°51.00' N, 130°01.00' W;
anchor (east shore):
53°51.96' N, 129°58.27' W (NAD 27)

Kumealon Island Cove Chart 3773;
entrance: 53°51.52' N, 130°01.73' W;
anchor: 53°51.74' N, 130°01.53' W (NAD 27)

Kumeon Bay Chart 3994;
anchor: 54°42.56' N, 130°14.57' W (NAD 27)

Kunakun Point Light Chart 3869;
position: 53°28.15' N, 132°53.72' W (NAD 27)

Kunechin Islets Light Chart 3512;
position: (S. tip of largest isl.):
49°37.20' N, 123°48.17' W (NAD 27)

"Kunechin Islets, North Cove"
Chart 3312;
anchor: 49°36.93' N, 123°48.04' W (NAD 27)

Kwakshua Channel Chart 3784;
east entrance: 51°39.00' N, 127°57.30' W;
north entrance: 51°42.30' N, 128°04.00' W
(NAD 27)

Kwakume Inlet Charts 3784 or 3727;
entrance: 51°42.45' N, 127°53.35' W;
anchor (south): 51°42.17' N, 127°52.66' W;
anchor (inner cove):
51°42.73' N, 127°51.57' W (NAD 27)

Kwakume Point
Chart 3784; point position:
51°41.57' N, 127° 53.23' W (NAD 27)

Kwakume Point Light Chart 3784;
position: (SW end of isl.):
51°41.55' N, 127°53.28' W (NAD 27)

Kwatna Bay Chart 3729
entrance: 52°06.72' N, 127°25.90' W
(NAD 27)

Kwatna Inlet Chart 3729;
entrance: 52°04.50' N, 127°28.00' W (NAD 27)

Kwatsi Bay ("Glory Be Basin")
Chart 3515;
anchor: 50°52.08' N, 126°15.05' W (NAD 83)

Kwinamass Bay Chart 3994;
position: 54°47.05' N, 130°10.04' W (NAD 27)

Kwuna Point Light Chart 3890;
position: (outer end of wharf):
53°12.92' N, 131°59.43' W (NAD 27)

Kxngeal Inlet Chart 3772;
entrance: 53°44.05' N, 129°49.45' W;
anchor (beach): 53°45.22' N, 129°49.35' W
(NAD 27)

Kynoch Inlet Chart 3962;
entrance: 52°45. 80' N, 128°07.15' W
(NAD 27)

**Kynumpt Harbour "Strom Bay
and Cove"** Charts 3787, 3720;
anchor (Strom Bay):
52°12.44' N, 128°09.79' W;
anchor (Strom Cove):
52°12.17' N, 128°09.90' W (NAD 27)

Kyuquot Bay Chart 3682;
anchor: 49°59.25' N, 127°17.47' W (NAD 27)

**Kyuquot Canyon ODAS Light
Buoy 46132** Chart 3604;
position: 49°43.90' N, 127°55.35' W
(NAD 27)

Kyuquot Channel Light Buoy M38
Chart 3682;
position: 49°56.13' N, 127°17.57' W
(NAD 27)

Kyuquot Sound
Charts 3682, 3683, 3623, 3651;
position: 50°00' N, 127°13' W (NAD 27)

Labouchere Channel Chart 3730;
south entrance: 52°21.00' N, 127°10.90' W;
north entrance: 52°26.80' N, 127°15.10' W
(NAD 27)

"Lacy Falls" Chart 3515;
position: 50°50.96' N, 126°19.50' W
(NAD 83)

Lady Trutch Passage Chart 3728;
north entrance: 52°20.96' N, 128°20.69' W;
anchor: 52°21.80' N, 128°20.50' W
(NAD 27)

**Ladysmith (Oyster Harbour),
Sibell Bay** Chart 3475;
public docks: 48°59.97' N, 123°48.76' W;
anchor (deep in harbor):
49°00.70' N, 123°49.65' W;
or anchor Sibell Bay:
48°59.47'N, 123°47.11' W (NAD 27);

Lagoon Cove Chart 3541;
anchor: 50°16.32' N, 124°43.57' W
(NAD 27)

Lagoon Cove Chart 3564;
float: 50°35.91' N, 126°18.83' W (NAD 83)

Lagoon Inlet Chart 3894;
entrance: 53°55.70' N, 131°54.30' W;
anchor (near pier):
53°55.68' N, 131°56.70' W;
anchor (lagoon): 52°56.10' N, 131°57.60' W
(NAD 27)

Lake Bay, Read Island 3539, 3538;
entrance: 50°08.30' N, 125°07.00' W
(NAD 83)

Lama Passage Chart 3785;
east entrance: 52°04.10' N, 127°56.70' W;
north entrance: 52°11.20' N, 128°06.00' W
(NAD 27)

Lancelot Inlet Charts 3559, 3312, 3538;
entrance: 50°02.20' N, 124°43.50' W
(NAD 27)

Langara Point Light Chart 3868;
position: (NW pt of Langara Isl.):
54°15.38' N, 133°03.50' W (NAD 27)

Langford Cove Chart 3728;
entrance: 52° 20.52' N, 128°36.90' W;
anchor (south nook):
52°20.29' N, 128°37.21' W (NAD 27)

"Langley Passage Anchorages"
Chart 3795;
anchor (0.3 mile southwest of **"ET4"**):
53°02.82' N, 129°37.71' W;
anchor Ethelda Bay:
53°03.33' N, 129°40.50' W (NAD 27)

**Laperouse Bank ODAS Light
Buoy 46206** Chart 3602;
position: 48°50.00' N, 126°00.00' W (NAD 27)

Larcom Lagoon Chart 3920;
entrance: 55°23.26' N, 129°45.14' W;
anchor: 55°23.87' N, 129°44.42' W (NAD 83)

Laredo Channel Charts 3737;
southeast entrance:
52°38.50' N, 128°53.80' W;
northwest entrance:
52°51.00' N, 129°12.00' W (NAD 27)

Laredo Inlet Chart 3737;
entrance: 52°37.40' N, 129°48.90' W (NAD 27)

Laredo Sound Charts 3726, 3728, 3737;
south entrance: 52°20.00' N, 128°50.00' W
(NAD 27)

"Larkin Point Basin" Charts 3726, 3737
(uncharted);
entrance: 52°29.80' N, 128°48.75' W;
anchor: 52°31.00' N, 128°48.80' W (NAD 27)

Larsen Harbour Charts 3747, 3927;
entrance (0.85 mile due west of White Rocks
Light): 53°38.10' N, 130°32.37' W;
inner entrance (100 yards west of QR light):
53°37.76' N, 130°32.43' W;
anchor (mud flat):
53°37.38' N, 130°33.00' W (NAD 27)

Larsen Harbour Light Chart 3747;
position: (westerly end of Banks Isl.):
53°37.75' N, 130°32.30' W (NAD 27)

Larso Bay Chart 3730;
anchor: 52°10.85' N, 126°51.58' W (NAD 27)

Lassiter Bay Chart 3552 (inset);
position: 51°08.60' N, 127°36.60' W (NAD 27)

Laura Bay North Chart 3515;
anchor: 50°49.47' N, 126°34.10' W (NAD 83)

Laura Cove Chart 3555;
anchor: 50°08.77' N, 124°40.10' W (NAD 27)

Laurel Point Light Chart 3415;
position: (NW extremity of pt):
48°25.48' N, 123°22.56' W (NAD 27)

Law Island Light Chart 3720;
position: (N. end of isl.):
52°16.08' N, 128°10.29' W (NAD 27)

Lawn Point Light and Bell Buoy C18
Chart 3890;
position: (SE of Lawn Pt light):
53°23.90' N, 131°53.80' W (NAD 27)

**Lawn Point Light and Whistle Buoy
C14** Chart 3890;
position: (off pt): 53°25.56' N, 131°53.13' W
(NAD 27)

Lawn Point Range Light Chart 3890;
position: (at pt): 53°25.50' N, 131°54.84' W
(NAD 27)

Lawson Harbour Chart 3927;
entrance: 54°01.83' N, 130°15.13' W;
anchor: 54°01.40' N, 130°15.00' W (NAD 27)

Lawyer Islands Charts 3956, 3717, 3927;
fairway 0.35 mile west of Genn Island light:
54°05.90' N, 130°18.22' W (NAD 83)

Lawyer Islands Light Chart 3957;
position: (on summit of northernmost isl., at
NW end): 54°06.77' N, 130°20.61' W
(NAD 83)

Leading Point Light Chart 3920;
position: 54°58.82' N, 129°50.92' W
(NAD 83)

Learmonth Island Light Chart 3893;
position: (on rock, N. of isl.):
53°40.72' N, 132°27.20' W (NAD 27)

Leask Creek
Chart 3542 (new 94 ed.), 3312;
position: 50°28.13' N, 125°02.75' W (NAD 83)

"Leckie Bay Cove" Chart 3784;
position: 51°48.50' N, 128°06.50' W (NAD 27)

Leeson Point Light Chart 3681;
position: (near S. extremity of reef off pt):
50°32.12' N, 127°37.59' W (NAD 83)

Legace Point Light Chart 3734;
position: (at the pt):
52°27.77' N, 128°25.10' W (NAD 27)

Legge Point Light Chart 3711;
position: (NW end of Cone Isl.):
52°36.23' N, 128°31.10' W (NAD 27)

Lemmens Inlet Charts: 3685, 3649
entrance: 49°10.30' N, 125°53.47' W
(NAD 27)

"Lemmens Northwest Cove"
Chart 3649;
anchor: 49°13.43' N, 125°51.50' W (NAD 27)

Lennard Island Chart 3685;
lighthouse: 49°06.64' N, 125°55.33' W
(NAD 83)

Lennard Island Light Chart 3685;
position: (on SW pt of isl.):
49°06.64' N, 127°55.33' W (NAD 83)

Leroy Bay Chart 3934;
position: 51°16.34' N, 127°39.90' W (NAD 83)

Levy Point Light Chart 3742;
position: (NE end of Ashdown Isl.):
53°04.68' N, 129°12.05' W (NAD 27)

Lewall Inlet Chart 3784;
entrance: 51°45.98' N, 128°04.74' W;
anchor: 51°46.11' N, 128°06.19' W (NAD 27)

Lewis Cove Chart 3547;
anchor: 50°49.40' N, 127°03.13' W (NAD 83)

Lewis Point Light Chart 3546;
position: (W. side of entrance to Beaver
Cove): 50°33.11' N, 126°51.26' W (NAD 83)

Lewis Reef Light Chart 3424;
position: (on reef): 48°25.53' N,
123°16.78' W (NAD 83)

Liddle Island Light Chart 3920;
position: (entrance to Alice Arm):
55°23.75' N, 129°41.48' W (NAD 83)

Light and Bell Buoy D75 Chart 3957;
position: 54°16.48' N, 130°47.50' W
(NAD 83)

Light Buoy S10 Chart 3490;
position: (S. side of channel):
49°07.78' N, 123°13.78' W (NAD 83)

Light Buoy S12 Chart 3490;
position: (S. side of channel):
49°07.74' N, 123°13.35' W (NAD 83)

Light Buoy S14 Chart 3490;
position: (S. side of channel):
49°07.65' N, 123°12.83' W (NAD 83)

Light Buoy S15 Chart 3490;
position: (N. side of channel):
49°07.69' N, 123°12.57' W (NAD 83)

Light Buoy S16 Chart 3490;
position: (at turn in New Cut channel):
49°07.22' N, 123°11.73' W (NAD 83)

Light Buoy S17 Chart 3490;
position: (N. side of channel):
49°07.33' N, 123°11.60' W (NAD 83)

Light Buoy S19 Chart 3490;
position: (S. of Steveston Bar):
49°06.98' N, 123°10.65' W (NAD 83)

Light Buoy S2 Chart 3490;
position: (on S. side of channel):
49°06.18' N, 123°17.95' W (NAD 83)

Light Buoy S21 Chart 3490;
position: 49°06.62' N, 123°09.20' W
(NAD 83)

Light Buoy S23 Chart 3490;
position: (N. side of channel):
49°06.98' N, 123°05.42' W (NAD 83)

Light Buoy S25 Chart 3490;
position: (opposite Deas Island):
49°07.30' N, 123°04.63' W (NAD 83)

Light Buoy S26 Chart 3490;
position: (S. side of channel):
49°07.53' N, 123°04.05' W (NAD 83)

Light Buoy S28 Chart 3490;
position: (N. of Deas Island):
49°08.38' N, 123°02.92' W (NAD 83)

Light Buoy S30 Chart 3490;
position: (NE of Deas Island):
49°08.73' N, 123°02.23' W (NAD 83)

Light Buoy S32 Chart 3490;
position: (S. side of channel):
49°09.32' N, 122°59.66' W (NAD 83)

Light Buoy S34 Chart 3490;
position: (opposite Purfleet Pt):
49°09.39' N, 122°58.85' W (NAD 83)

Light Buoy S4 Chart 3490;
position: (on S. side of channel):
49°06.68' N, 123°16.77' W (NAD 83)

Light Buoy S6 Chart 3490;
position: (on S. side of channel):
49°07.19' N, 123°15.58' W (NAD 83)

Light Buoy S8 Chart 3490;
position: (on S. side of channel):
49°07.71' N, 123°14.41' W (NAD 83)

Lighthouse Bay Chart 3313, p. 16;
position: 49°00.56' N, 123°35.14' W
(NAD 83)

"Lime Point Cove" Chart 3738;
anchor: 52°46.68' N, 128°20.10' W (NAD 27)

Limestone Bay Chart 3668, 646 (inset);
anchor: 48°59.12' N, 124°58.33' W (NAD 83)

Limestone Inlet Light Chart 3646;
position: (on islet):
48°58.95' N, 124°58.29' W (NAD 83)

**Lions Gate Bridge North Sector
Light** Chart 3493;
position: 49°18.96' N, 123°08.21' W
(NAD 83)

**Lions Gate Bridge South Sector
Light** Chart 3493;
position: 49°18.89' N, 123°08.26' W
(NAD 83)

Little Group Rock Light Chart 3476;
position: (centre of passage between Coal
and Ker Isl.): 48°40.58' N, 123°21.99' W
(NAD 27)

Little Nimmo Bay Chart 3547;
anchor: 50°56.30' N, 126°40.90' W (NAD 83)

Little River Charts 3527, 3513;
position: 49°44.43' N, 124°55.27' W
(NAD 27)

"Little Thompson Bay and Cove"
Chart 3787;
entrance: 52°09.00' N, 128°22.40' W;
anchor (Little Thompson Cove):
52°09.36' N, 128°20.77' W;
anchor ("The Nook"):
52°10.00' N, 128°21.03' W (NAD 27)

Little Zero Rock Light Buoy V30
Chart 3440;
position: (W. of rock):
48°31.92' N, 123°19.67' W (NAD 27)

Lizard Point Light Chart 3546;
position: (NE side, Malcolm Isl.):
50°40.29' N, 126°53.60' W (NAD 83)

Lizard Point Light Chart 3994;
position: (extremity of pt, Portland Inlet):
54°50.03' N, 130°16.46' W (NAD 27)

Lizzie Cove Chart 3785;
entrance; 52°04.90' N, 128°04.90' W;
position: 52°03.15' N, 128°05.20' W (NAD 27)

Lockhart Bay Chart 3787;
anchor: 52°12.28' N, 128° 15.95' W (NAD 27)

"Log Boom Cove" Chart 3312, 3512;
position 49°30.05' N, 124°12.56' N (NAD 27)

Logan Bay Chart 3746;
entrance: 53°35.82' N, 130°14.03' W
(NAD 27)

Logan Rock Light Chart 3724;
position: 53°02.25' N, 129°28.57' W
(NAD unknown)

Lombard Point Light Chart 3728;
position: 52°28.82' N, 128°56.53' W (NAD 27)

Lone Tree Point Light Chart 3668;
position: (on pt, E. side of Stamp Narrows):
49°11.05' N, 124°48.99' W (NAD 83)

Long Bay Chart 3312, 3512;
position: 49°29.95' N, 124°12.55' W (NAD 27)

Long Harbour Chart 3313, p. 10;
anchor: 48°51.64' N, 123°27.90' W (NAD 83)

Long Harbour Light Chart 3478;
position: (on rock): 48°51.12' N,
123°26.45' W (NAD 83)

Long Inlet Chart 3891;
entrance: 53°12.10' N, 132°14.00' W (NAD 83)

Long Point Cove Chart 3785;
entrance: 55°03.25' N, 127°56.90' W;
anchor: 52°02.80' N, 127°57.03' W (NAD 27)

Lookout Island Light Chart 3683;
position: 49°59.88' N, 127°26.86' W (NAD 27)

Lookout Point Light Chart 3481;
position: (on pt): 49°22.62' N, 123°17.34' W
(NAD 27)

Loquillilla Cove Chart 3549;
anchor: 50°51.67' N, 127°45.45' W (NAD 83)

"Loretta Island Cove" Chart 3743;
entrance: 53°43.30' N, 128°51.70' W;
anchor: 53°44.00' N 128 50.72' W (NAD 27)

Lorte Island (Cordero Lodge)
Chart 3543;
float: 50°26.83' N, 125°27.07' W (NAD 83)

Lorte Island Light Chart 3543;
position: (S. tip of isl.):
50°26.67' N, 125°27.17' W (NAD 83)

Loughborough Inlet Chart 3543;
entrance: 50°27.00' N, 125°36.30' W (NAD 83)

Louie Bay Chart 3663;
anchor: 49°44.77' N, 126°56.09' W (NAD 27)

Louisa Cove Chart 3711 (inset);
entrance: 52°11.60' N, 128°28.80' W;
anchor: 52°10.92' N, 128°29.12' W (NAD 27)

"Louise Channel South" Chart 3787;
entrance: 52°05.50' N, 128°22.80' W;
anchor: 52°06.58' N, 128°22.23' W (NAD 27)

Louise Narrows Chart 3894; north
entrance: 52°57.60' N, 131°54.15' W;
south entrance: 52°56.57' N, 131°54.04' W
(NAD 27)

Low Island Light Chart 3807;
position: (NW end of northernmost Low Isl.):
52°54.75' N, 131°32.30' W (NAD 27)

Lower Rapids Chart 3537;
position: 50°18.60' N, 125°15.80' W
(NAD 27)

Lowrie Bay Chart 3624;
entrance: 50°41.91' N, 128°21.10' W (NAD 27)

**Luard Shoal Light and Whistle Buoy
E63** Chart 3728;
position: 52°24.23' N, 128°53.47' W
(NAD 27)

Lucy Islands Light Chart 3957;
position: (NE side of E. isl.): 54°17.75' N,
130°36.52' W (NAD 83)

Lucy Islands North Light Chart 3957;
position: 54°18.08' N, 130°37.30' W
(NAD 83)

Lund Breakwater Centre Light
Chart 3311;
position: 49°58.84' N, 124°45.69' W
(NAD 83)

Lund Breakwater North Light
Chart 3311;
position: 49°58.86' N, 124°45.67' W
(NAD 83)

Lund Breakwater South Light
Chart 3311;
position: 49°58.83' N, 124°45.71' W
(NAD 83)

Lund Charts 3311, sheet 5, 3538;
north breakwater light:
49°58.86' N, 124°45.67' W (NAD 83)

Lund Light Chart 3311;
position: (E. end of S. Copeland Isl.):
49°59.76' N, 124°47.60' W (NAD 83)

Lundy Cove Chart 3721 (inset);
entrance: 53°24.95' N, 129°50.90' W;
anchor (outer bay):
53° 24.73' N, 129° 50.24' W;
anchor (east nook):
53°24.55' N, 129°49.74' W (NAD 27)

Lyall Harbour Chart 3313, p. 12;
anchor: 48°47.77' N, 123°10.87' W (NAD 83)

Lyall Island Light Chart 3543;
position: (SW pt of isl.):
50°26.70' N, 125°35.65' W (NAD 83)

"Lyall Point Bight" Chart 3670;
anchor: 48°58.54' N, 125°17.42' W (NAD 83)

Lyall Point Light Chart 3670;
position: 48°58.17' N, 125°19.34' W
(NAD 83)

Lyell Bay Chart 3808;
entrance: 52°39.45' N, 131°39.85' W;
anchor 52°38.95' N, 131°38.70' W (NAD 27)

MacDonald Bay Chart 3742;
entrance: 53°11.82' N, 129°20.64' W;
anchor (east end):
53°11.82' N, 129°19.45' W (NAD 27)

Macdonald Island Chart 3312;
position: 50°11.20' N, 123°48.17' W
(NAD 27)

Mackenzie Sound Chart 3547;
entrance: 50°56.00' N, 126°46.50' W
(NAD 83)

Macktush Creek Light Chart 3668;
position: (at mouth of creek):
49°06.57' N, 124°49.35' W (NAD 83)

Madeira Park, Welbourn Cove
Chart 3535;
public float: 49°37.42' N, 124°01.44' W
(NAD 27)

Madrona Bay Chart 3313, p. 10;
anchor: 48°51.46' N, 123°29.17' W (NAD 83)

Magee Channel Chart 3934;
west entrance: 51°29.52' N, 127°40.95' W;
anchor: 51°29.36' N, 127°38.72' W (NAD 83)

"Magin Islets Cove" Chart 3547;
anchor: 50°52.45' N, 126°39.60' W (NAD 83)

Mahatta Creek Chart 3679;
anchor: 50°27.63' N, 127°51.70' W (NAD 83)

**Maitland Island Southwest End
Light** Chart 3743;
position: 53°42.15' N, 129°04.47' W (NAD 27)

Major Brown Rock Light Chart 3934;
position: (on rock, entrance to Rivers Inlet):
51°25.46' N, 127°42.03' W (NAD 83)

Major Islet Light Chart 3311;
position: (S. end of isl.):
49°59.33' N, 124°48.90' W (NAD 83)

Makwazniht Island Light Chart 3681;
position: (on isl., Quatsino Narrows):
50°33.37' N, 127°33.32' W (NAD 83)

Malaspina Inlet Charts 3559, 3312, 3538;
entrance: 49°04.30' N, 124°48.60' W (NAD 27)

Malibu Rapids Chart 3312;
entrance: 50°09.72' N, 123°51.92' W
(NAD 27)

Malibu Rapids Light Chart 3514;
position: (entrance to rapids):
50°09.73' N, 123°51.18' W (NAD 83)

Malksope Inlet Chart 3683
entrance: 50°05.70' N, 127°29.40' W
(NAD 27)

Mamalilaculla, Village Island
Charts 3545, 3546;
anchor: 50°37.44' N, 126°34.72' W (NAD 83)

**Mandarte Island East Cardinal Light
Buoy UT** Chart 3441;
position: 48°37.57' N, 123°15.92' W (NAD 27)

Mandarte Island North Light
Chart 3441;
position: 48°38.31' N, 123°17.78' W (NAD 27)

Mannion Bay (Deep Bay)
Chart 3534 (inset);
position: 49°22.96' N, 123°19.76' W (NAD 83)

Manson Bay Charts 3538, 3311;
public float: 50°04.30' N, 124°59.00' W
(NAD 83)

Mantrap Inlet Chart 3921;
narrows entrance:
51°35.53' N, 127°45.36' W;
anchor (north): 51°35.84' N, 127°44.78' W;
anchor (south): 51°35.04' N, 127°45.41' W
(NAD 83)

Manzanita Cove Chart 3994;
entrance: 54°45.65' N, 130°26.10' W;
anchor: 54°45.44' N, 130°26.18' W (NAD 27)

Maple Bay (Birds Eye Cove)
Chart 3313, p. 14;
public floats: 48°48.90'N, 123°36.55' W;
anchor Birds Eye Cove:
48°47.82' N, 123°35.98' W (NAD 83)

Maple Bay Chart 3393;
entrance: 55°25.40' N, 130°00.90' W;
position: 55°25.20' N, 130°00.55' W
(NAD 27)

**Maple Bay Speed Control Light
Buoy** Chart 3478;
position: 48°48.40' N, 123°35.75' W
(NAD 83)

Maple Spit Light Buoy P39
Chart 3527;
position: 49°28.50' N, 124°43.40' W (NAD 27)

Marble Cove Chart 3668, 3671;
anchor: 48°54.76' N, 125°06.69' W (NAD 83)

Marble Island Light Chart 3891;
position: (W. end of isl.):
53°12.05' N, 132°40.10' W (NAD 83)

Marcus Passage Chart 3717;
north entrance (0.45 mile NW Hazel Point.):
54°07.20' N, 130°15.40' W (NAD 27)

Margaret Bay Chart 3313, p. 5;
anchor: 48°29.80' N, 123°18.60' W (NAD 83)

Margaret Bay Chart 3931;
entrance: 51°19.89' N, 127°31.29' W;
anchor: 51°20.00' N, 127°29.62' W (NAD 83)

Marked Tee Bluff Light Chart 3717;
position: (northern end of Kennedy Isl.):
54°04.15' N, 130°09.80' W (NAD 83)

Markle Inlet Chart 3746;
entrance: 53°34.25' N, 129°57.12' W (NAD 27)

Markle Passage Chart 3746;
west entrance: 53°33.65' N, 129°59.75' W;
anchor (island (325) bight):
53°34.38' N, 129°57.82' W (NAD 27)

Marktosis Light Chart 3643;
position: (in shallow bay at entrance to
Marktosis): 49°16.71' N, 126°03.62' W
(NAD unknown)

Marmot Bay Charts 3794, 3933;
position: 55°53.08' N, 130°00.33' W
(NAD 27)

Marsh Bay Chart 3548 ;
position: 50°55.35' N, 127°22.00' W (NAD 83)

Marshall Inlet Chart 3808;
entrance: 52°28.70' N, 131°27.90' W;
anchor: 52°28.00' N, 131°30.36' W (NAD 27)

Martin Cove and Francis Cove
Chart 3535 (inset);
position: Francis Cove,
49°36.64' N, 124°03.55' W;
Martin Cove, 49°37.06' N, 124°03.73' W
(NAD 27)

Martin Islets Chart 3545;
anchor: 50°39.47' N, 126°14.33' W (NAD 83)

Martins Cove Charts 3785, 3787;
floats: 52°10.32' N, 128°08.38' W (NAD 27)

Marvinas Bay Chart 3664;
anchor 49°39.54' N, 126°37.37' W (NAD 27)

Mary Anne Point Light Chart 3473;
position: (on the pt):
48°51.73' N, 123°18.73' W (NAD 27)

Mary Basin Chart 3663;
anchor: 49°46.86' N, 126°50.10' W (NAD 27)

Mary Cove Chart 3734;
entrance: 52°36.60' N, 128°26.65' W;
anchor: 52°36.87' N, 128°26.16' W (NAD 27)

Mary Point Light Chart 3743;
position: (on pt, Verney Passage):
53°33.00' N, 128°58.00' W (NAD 27)

Mary Tod Island Light Chart 3424;
position: (SW extremity of breakwater):
48°25.55' N, 123°17.92' W (NAD 83)

**Masset Harbour Entrance Light
Buoy C31** Chart 3895;
position: (NW of Entry Point):
54°03.73' N, 132°12.40' W (NAD 27)

**Masset Harbour Entrance Range
Light** Chart 3895;
position: (NW of Entry Point):
54°01.98' N, 132°12.40' W (NAD 27)

Masset Harbour Light and Bell Buoy C29 Chart 3895;
position: (W. of Outer Bar):
54°05.75' N, 132°12.87' W (NAD 27)

Masterman Islands Light Chart 3550;
position: (NE extremity of northeasterly island group): 50°45.52' N, 127°25.30' W (NAD 83)

Mate Island Light Chart 3643;
position: (E. end of isl.):
49°21.00' N, 126°15.93' W (NAD XX)

"Math Islands" Chart 3746;
position: 53°33.50' N, 129°57.00' W (NAD 27)

Matheson Inlet Chart 3808;
entrance: 52°28.40' N, 131°27.90' W;
anchor: 52°27.06' N, 131°28.51' W (NAD 27)

Mathieson Channel
Charts 3728, 3734, 3962;
south entrance: 52°18.20' N, 128°25.40' W;
north entrance: 52°50.70' N, 128°08.60' W (NAD 27)

Mathieson Narrows Chart 3962;
position: 52°50.70' N, 128°08.60' W (NAD 27)

Matilda Inlet Chart 3643
entrance: 49°18.13' N, 126°04.22' W (NAD unknown)

Matilda Inlet Light Chart 3643;
position: (on outside edge of reef, W. side of creek entrance): 49°18.21' N, 126°04.35' W (NAD unknown)

Matilpi Charts 3564, 3545;
anchor: 50°33.55' N, 126°11.30' W (NAD 83)

Matlset Narrows Chart 3649;
entrance: 49°14' N, 125°48' W (NAD 27)

"Matlset Narrows Cove" Chart 3649;
anchor: 49°13.98' N, 125°47.38' W (NAD 27)

Maud Island Cove Chart 3539 (inset);
position: 50°08.14' N, 125°20.63' W (NAD 83)

Maud Island Light Chart 3539;
position: (on rock, W. side of isl., Seymour Narrows): 50°07.83' N, 125°20.87' W (NAD 83)

Maud Island South Light Chart 3539;
position: 50°07.68' N, 125°20.54' W (NAD 83)

Maude Island (Nanoose Harbour) Light Chart 3459;
position: (near E. end of isl., N. side of entrance to hbr): 49°16.23' N, 124°04.60' W (NAD 27)

Maude Island Chart 3459;
position: 49°16.23' N, 124°04.60' W (NAD 27)

Maunsell Bay Chart 3552;
entrance: 51°05.00' N, 126°51.75' W;
anchor: 51°05.86' N, 126°47.54' W (NAD 27)

Maurus Channel Light Chart 3649;
position: (on rock, E. side of channel):
49°12.23' N, 125°55.77' W (NAD 27)

Mayne Passage (Blind Channel) Chart 3543;
north entrance: 50°25.90' N, 125°29.50' W;
south entrance; 50°22.90' N, 125°34.00' W (NAD 27)

"Maze Cove" Chart 3786;
anchor: 52°00.97' N, 128°12.73' W (NAD 27)

McBride Bay Chart 3663;
entrance 49°51' N, 126°44' W (NAD 27)

McBride Bay Chart 3931;
position: 51°18.47' N, 127°32.30' W (NAD 83)

McBride Bay Light Chart 3663;
position: 49°51.67' N, 126°43.70' W (NAD 27)

McClusky Bay Chart 3921;
anchor: 51°38.65' N, 127°47.43' W (NAD 83)

McCoy Cove Chart 3894;
entrance: 53°02.25' N, 131°39.55' W (NAD 27)

McCoy Cove Sector Light Chart 3894;
position: (E. side of cove):
53°02.10' N, 131°39.21' W (NAD 27)

McCreigh Point Light Chart 3742;
position: (SE of Pitt Isl.):
53°12.50' N, 129°30.00' W (NAD 27)

McEchran Cove Chart 3807;
entrance: 52°42.85' N, 131°48.80' W (NAD 27)

McEwan Rock Light Chart 3550;
position: (on rock):
51°03.53' N, 127°37.78' W (NAD 83)

McInnes Island Light Chart 3728;
position: (on bluff, S. side of isl., N. entrance
to Milbanke Sound):
52°15.70' N, 128°43.22' W (NAD 27)

McInnes Island Light Station
Chart 3728;
light: 52°15.70' N, 128°43.22' W (NAD 27)

McIntosh Bay Chart 3515 (inset);
anchor east McIntosh Bay:
50°51.77' N, 126°31.20' W (NAD 83)

"McIntosh Bay" Charts: 3648, 3649;
anchor: 49°12.55' N, 125°56.44' W (NAD 27)

McKay Bay Charts 3781, 3729;
entrance: 52°24.10' N, 127°26.40' W;
position: 52°24.77' N, 127°26.05' W (NAD 27)

McKay Island Light Chart 3648;
position: (W. side of isl.):
49°18.67' N, 126°03.45' W (NAD 27)

McKay Reach Chart 3742;
east entrance: 53°18.60' N, 128°55.00' W;
West entrance: 53°18.10' N, 129°06.30' W
(NAD 27)

McKinnon Lagoon Chart 3552;
entrance: 50°59.88' N, 127°13.95' W (NAD 27)

McLean Cove Chart 3682;
anchor: 49°58.3' N, 127°14.1' W (NAD 27)

McLeod Bay Chart 3544;
anchor west side: 50°28.22' N,
125°58.70' W (NAD 83)

McLoughlin Bay Charts 3785, 3787;
entrance: 52°08.40' N, 128°08.20' W;
position: 52°08.28' N, 128°08.47' W (NAD 27)

McMicking Inlet
Charts 3719 (inset), 3724, 3742;
inner entrance: 53°02.59' N, 129°26.98' W;
anchor (east cove):
53°04.87' N, 129°27.89' W (NAD 27)

McMullen Point Light Chart 3539;
position: (on pt): 50°14.76' N, 125°23.73' W
(NAD 83)

McMurray Bay Chart 3312;
position: 49°58.18' N, 124°00.43' W
(NAD 27)

"McNaughton Group Anchorage"
Chart 3786;
entrance:
51°54.60' N, 128°13.45' W (NAD 27)
1. **Bombproof Anchorage:**
Anchor in 5 fathoms,
51°55.46' N, 128°14.29' W
2. **Great Salt Lake Anchorage:**
Anchor in 7 fathoms,
51°56.29' N, 128°13.77' W
3. **Back Door Anchorage:**
Anchor in 10 fathoms,
51°55.97' N, 128°12.98' W
4. **Intersection Anchorage:**
Anchor in 14 fathoms,
51°55.98' N, 128°13.60' W
5. **Deep Anchorage:**
Anchor in 24 fathoms,
51°55.50' N, 128°13.20' W

McNiffe Rock Light Chart 3681;
position: (Neroutsos Inlet):
50°31.29' N, 127°36.75' W (NAD 83)

McRae Cove Chart 3311, sheet 4;
position: 49°44.76' N, 124°16.68' W
(NAD 83)

McRae Cove Chart 3734;
anchor: 52° 38.92' N, 128° 34.80' W (NAD 27)

Meade Bay Chart 3546;
anchor: 50°42.33' N, 126°35.23' W (NAD 83)

Meares Spit Light Buoy Y8 Chart 3685;
position: (W. end, Heynen Channel):
49°10.37' N, 125°55.87' W (NAD 83)

Megin River Chart 3648;
anchor: 49°26.13' N, 126°05.00' W (NAD 27)

Melanie Cove Chart 3555;
anchor: 50°08.53' N, 124°40.36' W (NAD 27)

Mellis Inlet Chart 3737;
entrance: 52°53.45' N, 128°43.00' W (NAD 27)

Menzies Bay Chart 3539 (inset);
position log booms, north side:
50°07.94' N, 125°22.44' W (NAD 83)

Mereworth Sound Chart 3552;
entrance: 51°08.10' N, 127°25.00' W
(NAD 27)

Mermaid Bay Chart 3543 (inset);
position: 50°24.22' N, 125°11.05' W
(NAD 27)

Merry Island Light Chart 3535;
position: (SE extremity of isl.,
SE entrance to Welcome Passage):
49°28.06' N, 123°54.67' W (NAD 27)

Metcalf Islands Bight Chart 3537;
position: 50°17.00' N, 125°21.80' W
(NAD 27)

Metlakatla Sector Light Chart 3955;
position: (N. tip of small isl., Venn Passage):
54°20.18' N, 130°25.33' W (NAD 27)

Meyers Narrows Chart 3710 (inset);
east entrance: 52°36.22' N, 128°35.72' W;
west entrance: 52°36.56' N, 128°37.88' W
(NAD 27)

"Meyers Narrows Cove"
Chart 3710 (inset);
anchor: 52°36.27' N, 128°37.00' W (NAD 27)

Meyers Passage
Charts 3734, 3710 (inset);
east entrance (Tolmie Channel):
52°40.20' N, 128°34.00' W;
west entrance: 52°35.88' N, 128°45.30' W;
anchor (elbow): 52°36.15' N, 128°35.22' W
(NAD 27)

"Middle Cove" Chart 3921;
anchor: 51°37.27' N, 127°45.70' W (NAD 83)

**Middle Nomad ODAS Light Buoy
46004** Chart 3000;
position: 50°58.00' N, 135°48.00' W
(NAD 27)

Middle Reef Light Buoy M41
Chart 3663;
position: (E. of reef):
49°48.10' N, 127°02.30' W (NAD 27)

Milbanke Sound Chart 3728;
south entrance: 52°12.00' N, 128°38.00' W;
north entrance: 52°25.00' N, 128°29.00' W
(NAD 27)

Miles Inlet Chart 3551;
entrance: 51°03.58' N, 127°36.31' W;
anchor (north arm):
51°04.03' N, 127°34.82' W (NAD 83)

Mill Bay Chart 3313, p.13;
position: 48°38.90' N, 123°32.95' W (NAD 83)

Mill Bay Chart 3920;
anchor (Mill Bay): 54°59.60′ N, 129°53.63′ W
(NAD 83)

Mill Bay Light Chart 3920;
position: 54°59.67′ N, 129°53.30′ W (NAD 83)

Millbrook Cove Chart 3934;
entrance: 51°19.14′ N, 127°43.96′ W;
anchor: 51°19.67′ N, 127°44.20′ W (NAD 83)

Miller Bay Chart 3958;
entrance: 54°16.03′ N, 130°15.56′ W (NAD 27)

Miller Inlet Chart 3721 (inset);
entrance: 53°28.18′ N, 129°54.63′ W;
anchor (inner basin south):
53° 28.43′ N, 129°52.22′ W (NAD 27)

Millstone Creek Light Buoy P11
Chart 3457;
position: 49°10.55′ N, 123°56.16′ W
(NAD 27)

Millstone Light Buoy P9 Chart 3457;
position: 49°10.41′ N, 123°56.03′ W
(NAD 27)

Miners Bay, Mayne Island
Chart 3313, p. 9;
anchor: 48°51.23′ N, 123°17.98′ W (NAD 83)

Minette Bay Chart 3736 ;
entrance: 53°59.40′ N, 128°39.65′ W;
position (float): 54°01.60′ N, 128°36.67′ W
(NAD 83)

"Mink Island Cove" Chart 3538;
anchor: 50°06.35′ N, 124°45.37′ W (NAD 27)

Mink Trap Bay, Burns Bay Chart 3721;
entrance: 53°26.50′ N, 129°51.60′ W
((NAD 27)

Minnis Bay Chart 3722;
entrance: 53°19.30′ N, 129°27.27′ W;
anchor: 53°25.52′ N, 129°27.78′ W (NAD 27)

Minstrel Island Charts 3564, 3545;
marina float: 50°36.80′ N, 126°18.30′ W
(NAD 83)

Misery Bay Chart 3311;
position: 49°40.88′ N, 123°34.20′ W (NAD 27)

Miskatla Inlet Chart 3743;
entrance and anchor (fishermen's Site):
53°47.85′ N, 128°57.20′ W;
anchor (inlet head):
53°51.05′ N, 128°55.10′ W (NAD 27)

Mist Rock Light Chart 3681;
position: 50°25.78′ N, 127°30.06′ W
(NAD 83)

Mitchell Bay Chart 3546;
public float: 50°37.83′ N, 126°51.09′ W;
anchor: 50°37.94′ N, 126°51.49′ W (NAD 83)

Mitchell Cove Chart 3722;
entrance: 53°21.76′ N, 129°28.14′ W;
anchor (two-snag island):
53°21.79′ N, 129°28.39′ W (NAD 27)

Mitlenatch Island Nature Park
Chart 3538;
entrance: 49°57.00′ N, 124°59.45′ W
anchor: 49°57.07′ N, 124°59.75′ W (NAD 27)

MK Bay Marina Chart 3736 (inset);
entrance: 53°59.04′ N, 128°39.27′ W (NAD 83)

Monckton Inlet Charts 3721 (inset), 3742;
entrance: 53°18.60′ N, 129°41.45′ W;
anchor (lagoon): 53°18.98′ N, 129°39.76′ W;
anchor (0.7 mile north of Roy Island):
53°19.76′ N, 129°36.83′ W;
anchor (head of inlet):
53°19.08′ N, 129°34.53′ W (NAD 27)

Monday Anchorage
Charts 3547, 3546, 3515;
anchor: 50°44.13′ N, 126°38.41′ W (NAD 83)
(The anchor coordinates are given for the
second island on the north side.)

Money Point Light Chart 3740;
position: (on pt) 53°22.92' N, 129°09.83' W
(NAD 27)

Monk Bay Chart 3737 uncharted;
entrance: 52°33.90' N, 128°52.10' W
(NAD 27)

Monks Islet Light Chart 3648;
position: (on islet, W. entrance to Calmus
Passage): 49°13.94' N, 126°00.90' W
(NAD 27)

Montague Harbour
Chart 3313, p. 9 (inset);
float: 48°53.83' N, 123°24.16' W;
anchor: 48°53.75' N, 123°24.00' W (NAD 83)

Moolock Cove Chart 3721 (inset);
entrance: 53°27.20' N, 129°50.25' W;
anchor (bight): 53°27.12' N, 129°49.16' W
(NAD 27)

Moon Bay Marina Chart 3736;
entrance: 53°59.26' N, 128°41.77' W
(NAD 83)

Moore Bay Chart 3515;
anchor east: 50°52.40' N, 126°32.50' W;
anchor west: 50°52.25' N, 126°33.95' W
(NAD 83)

Moore Cove Chart 3927;
position: 54°00.15' N, 130°05.70' W (NAD 27)

Moore Island Light Chart 3747;
position: (NW corner of isl.):
53°47.38' N, 130°31.22' W (NAD 27)

Mooyah Bay Chart 3664;
entrance: 49°38.20' N, 126°27.00' W
(NAD 27)

Morehouse Bay Chart 3940;
entrance: 52°17.40' N, 128°06.25' W;
anchor: 52°16.28' N, 128°05.10' W,
(NAD 83)

**Morehouse Passage and Lapwing
Island** Chart 3785 (inset);
Morehouse (south entrance):
51°51.30' N, 127°53.50' W (NAD 27)

**Morehouse Rock Bifurcation Light
and Bell Buoy EC** Chart 3737;
position: (SE of rock):
52°45.60' N, 129°05.90' W (NAD 27)

Moresby Camp Dock, Gordon Cove
Chart 3894;
entrance: 53°02.90' N, 132°00.90' W;
anchor: 53°02.50' N, 132°01.50' W (NAD 27)

**Moresby Island West ODAS
Light Buoy 46208** Chart 3002;
position: 52°30.00' N, 132°42.00' W (NAD 27)

Morfee Island Light Chart 3648;
position: (on SE. end of isl.):
49°13.08' N, 125°57.25' W (NAD 27)

Morgan Bay Chart 3934;
position: 51°31.96' N, 127°40.90' W (NAD 83)

Morris Bay Chart 3728;
entrance: 52°21.15' N, 128°26.85' W;
anchor: 52°20.89' N, 128°26.74' W (NAD 27)

Morse Basin Chart 3958;
position: 54°16.00' N, 130°15.00' W (NAD 83)

Moses Inlet Chart 3932;
entrance: 51°39.50' N, 127°27.50' W (NAD 83)

Mosquito Bay Chart 3940;
position: 52°23.70' N, 128°10.10' W (NAD 83)

Mosquito Harbour Chart 3649;
anchor: 49°13.05' N, 125°47.66' W (NAD 27)

Moss Passage, Sloop Narrows
Charts 3710, 3728;
east entrance: 52°21.68' N, 128°22.70' W;
west entrance: 52°21.30' N, 128°28.60' W
(NAD 27)

Mouat Bay Chart 3513;
launching ramp: 49°39.19' N, 124°27.70' W
(NAD 27)

Mouat Cove Chart 3728;
anchor: 52°16.57' N, 128°19.32' W (NAD 27)

Mouat Point Light Chart 3441;
position: (W. side of N. Pender Island):
48°46.21' N, 123°18.73' W (NAD 27)

Moulds Bay Charts 3539, 3538, 3312,;
anchor: 50°08.37' N, 125°11.52' W (NAD 83)

Moutcha Bay Chart 3664;
anchor: 49°47.00' N, 126°26.87' W (NAD 27)

"Mouth of Bay," Bullock Channel
Chart 3940;
anchor: 52°23.68' N, 128°04.54' W (NAD 83)

"Mt. Seafield Cove" Chart 3535 (inset);
position: 49°29.06' N, 123°57.44' W (NAD 27)

Muchalat Inlet Chart 3664
entrance: 49°39.05' N, 126°27.30' W
(NAD 27)

Muchalat Inlet East Light Chart 3664;
position: (on N. shore of inlet):
49°39.13' N, 126°15.67' W (NAD 27)

Muchalat Inlet Light Chart 3664;
position: (on pt, on N. shore of inlet):
49°38.72' N, 126°20.75' W (NAD 27)

Muchalat Inlet South Shore Light
Chart 3664;
position: (S. side of inlet):
49°39.25' N, 126°12.77' W (NAD 27)

Muir Cove Charts 3728, 3726;
position: 52°17.75' N, 128°39.50' W (NAD 27)

Murchison Island Chart 3808;
entrance: 52°35.90' N, 131°27.60' W;
buoys: 52°35.63' N, 131°27.96' W (NAD 27)

Murder Bay Charts 3641, 3440, 3430;
anchor: 48°20.00' N, 123°37.68' W (NAD 27)

Murder Cove Charts 3747, 3927;
anchor: 53°44.26' N, 130°19.44' W (NAD 27)

Murray Anchorage Chart 3724;
position: 53°00.78' N, 129°39.40' W (NAD 27)

"Murray Island Anchorage"
Chart 3544;
anchor: 50°29.77' N, 125°49.23' W (NAD 83)

Murray Labyrinth Chart 3921;
entrance: 51°02.59' N, 127°32.38' W;
anchor: 51°02.81' N, 127°31.88' W (NAD 83)

Mussel Bay Chart 3962;
position: 52°54.80' N, 128°02.10' W (NAD 27)

Mussel Inlet Chart 3962;
entrance: 52°51.00' N, 128°09.20' W
(NAD 27)

Mustang Bay Chart 3784;
position: 51°49.54' N, 128°02.50' W
(NAD 27)

**Mystery Reef Light and Bell Buoy
Q25** Chart 3311;
position: (NE of the reef):
49°54.77' N, 124°42.72' W (NAD 83)

Nabannah Bay Chart 3772;
entrance (north):
53°40.75' N, 129°46.00' W;
anchor (southeast corner):
53°40.30' N, 129°45.34' W (NAD 27)

Nahwitti Bar Chart 3549;
west entrance buoy:
50°54.15' N, 128°02.52' W (NAD 83)

**Nahwitti Bar Light and
Whistle Buoy MA** Chart 3549;
position: (W. of Nahwitti Pt):
50°54.08' N, 128°02.53' W (NAD 83)

Nahwitti Point Light Chart 3549;
position: (SW side of Hope Isl.):
50°54.31' N, 127°59.02' W (NAD 83)

Nakwakto Rapids Chart 3921;
position (Turret Rock):
51°05.79' N, 127°30.18' W;
north entrance: 51°06.00' N, 127°30.20' W
(NAD 83)

"Nalau Inlet" Chart 3784;
entrance: 51°44.58' N, 128°02.74' W (NAD 27)

Nalau Passage Chart 3784;
west entrance: 51°47.10' N, 128°07.30' W;
east entrance: 51°48.40' N, 128°00.70' W
(NAD 27)

Namu Harbour Chart 3785 (inset), 3784;
south entrance: 51°51.30' N, 127°53.50' W
north float: 51°51.73' N, 127°51.93' W
(NAD 27)

Nanaimo Harbour Chart 3313, p. 21;
floating breakwater:
49°10.20' N, 123°55.95' W (NAD 83)

**Nanaimo Harbour Entrance Groyne
Light** Chart 3457;
position: (E. of Assembly wharf):
49°09.82' N, 123°55.18' W (NAD 27)

Nanaimo Harbour Entrance Light
Chart 3457;
position: (S. side of entrance):
49°09.95' N, 123°54.99' W (NAD 27)

**Nanakwa Shoal ODAS Light Buoy
46181** Chart 3743;
position: 53°50.03' N, 128°49.92' W (NAD 27)

Nanoose Harbour Chart 3459;
anchor: 49°15.27' N, 124°07.83' W (NAD 27)

Nanoose Harbour Light Chart 3459;
position: (on pt SW of Richard Point):
49°15.93' N, 124°07.40' W (NAD 27)

Napier Point Light Chart 3787;
position: (on the pt):
52°07.93' N, 128°07.95' W (NAD 27)

"Narrows Cove" Chart 3921;
entrance: 51°36.38' N, 127°44.16' W
(NAD 83)

Narvaez Bay Chart 3313, p. 24;
anchor: 48°46.48' N, 123°06.04' W (NAD 83)

Nascall Bay Charts 3729, 3730;
anchor: 52°29.63' N, 127°16.43' W
(NAD 27)

Nascall Island and Rocks
Charts 3729, 3730;
position: 52°30.60' N, 127°14.70' W
(NAD 27)

**"Nash Narrows Cove,"
Spiller Channel** Chart 3940;
anchor: 52°30.45' N, 128°01.96' W
(NAD 83)

"Nash Passage" Chart 3940;
south entrance: 52°27.90' N, 128°04.00' W;
north entrance: 52°30.40' N, 128°01.60' W
(NAD 83)

Nasoga Gulf Chart 3933;
entrance: 54°49.70' N, 130°10.00' W;
anchor: 54°53.70' N, 130°04.00' W (NAD 27)

Nasparti Inlet Chart 3683
entrance: 50°06.30' N, 127°40.30' W
(NAD 27)

Nass Harbour Chart 3920;
position: 54°56.30' N, 129°56.10' W (NAD 83)

"Nass Point Cove" Chart 3933;
anchor: 55°03.10' N, 129°59.00' W (NAD 27)

Nass River Chart 3920;
entrance: 54°59.30' N, 130°00.50' W (NAD 83)

Naysash Bay Chart 3931;
position: 51°19.00′ N, 127°20.55′ W (NAD 83)

Naysash Inlet Chart 3931;
entrance: 51°18.37′ N, 127°22.10′ W (NAD 83)

Neekas Cove Chart 3940;
entrance: 52°26.75′ N, 128°09.30′ W;
anchor: 52°27.99′ N, 128°09.57′ W (NAD 83)

Neekas Inlet Chart 3940;
entrance: 52°26.10′ N, 128°08.70′ W;
anchor: 52°27.77′ N, 128°12.32′ W (NAD 83)

**Neill Ledge Bifurcation Light
Buoy NP** Chart 3546;
position: (E. extremity of ledge):
50°36.16′ N, 127°02.29′ W (NAD 83)

Neill Rock Light Buoy N21 Chart 3546;
position: (N. side of rock):
50°36.58′ N, 127°03.18′ W (NAD 83)

Nels Bight Chart 3624;
entrance: 50°48.03′ N, 128°22.50′ W (NAD 27)

Nelson Road Light Chart 3490;
position: (N. bank of river):
49°09.13′ N, 123°01.30′ W (NAD 83)

Nelson Rock Light Chart 3311;
position: (on rock):
49°38.65′ N, 124°07.07′ W (NAD 83)

Nenahlmai Lagoon Chart 3552;
entrance: 50°59.70′ N, 127°14.20′ W (NAD 27)

Nepean Rock Light Buoy E78
Chart 3742;
position: 53°12.25′ N, 129°36.33′ W (NAD 27)

Nepean Sound Chart 3742
position: 53°13.50′ N, 129°41.00′ W (NAD 27)

Neroutsos Inlet Chart 3679 and 3681
entrance: 50°30.00′ N, 127°35.00′ W
(NAD 83)

Nesook Bay Chart 3664;
anchor: 49°46.17′ N, 126°24.24′ W (NAD 27)

**Nettle Basin, Lowe Inlet
(Verney Falls)** Chart 3772;
entrance: 53°32.50′ N, 129°35.85′ W;
anchor (Verney Falls):
53°33.60′ N, 129°33.93′ W;
anchor (Nettle Basin):
53°33.43′ N, 129°34.06′ W (NAD 27)

Nettle Island Chart 3670;
anchor: 48°55.75′ N, 125°14.97′ W (NAD 83)

New Brighton Chart 3526;
public float: 49°27.01′ N, 123°26.36′ W
(NAD 27)

New Cut Range 1 Light Chart 3490;
position: 49°07.92′ N, 123°13.28′ W
(NAD 83)

New Cut Range 2 Light Chart 3490;
position: 49°06.62′ N, 123°10.02′ W
(NAD 83)

**New Westminster Railway Swing
Bridge Light** Chart 3490;
position: 49°12.52′ N, 122°53.60′ W
(NAD 83)

**Newcastle Island Provincial Park
(Mark Bay)** Chart 3313, p. 21;
anchor: 49°10.85′ N, 123°55.93′ W (NAD 83)

Newcombe Harbour Chart 3753 (inset);
entrance: 53°41.82′ N, 130°06.11′ W;
anchor (east nook):
53°42.27′ N, 130°05.52′ W,
anchor (north): 53°42.86′ N, 130°05.08′ W
(NAD 27)

Newton Cove Chart 3663;
anchor: 49°52.49′ N, 126°56.46′ W (NAD 27)

"NFG" Cove Chart 3784;
entrance: 51°50.31′ N, 128°06.62′ W (NAD 27)

Nicholson Cove Chart 3313, p. 18;
position: 49°02.90' N, 123°45.45' W
(NAD 83)

Nicomekl River Chart 3463;
position: 49°04.40' N, 122°50.10' W
(NAD 27)

"Nigei Island East Cove" Chart 3549;
anchor: 50°52.13' N, 127°40.37' W (NAD 83)

Nimpkish Bank Light Chart 3546;
position: (on shoal):
50°34.74' N, 126°56.88' W (NAD 83)

Nimpkish River Chart 3546;
anchor: 50°34.30' N, 126°57.60' W (NAD 83)

"9 Fathom Cove" Chart 3312;
anchor: 49°43.05' N, 124°12.41' W (NAD 27)

Nissen Bight Charts 3598 or 3624;
entrance: 50°48.50' N, 128°18.30' W (NAD 27)

Nitinat Narrows and Nitinat Lake
Chart 3647, 3606;
Nitinat Bar: 48°40.11' N, 124°51.05' W
(NAD 27)

"No Name Bay" Chart 3552;
position: 51°09.25' N, 127°06.30' W
(NAD 27)

"No Name Cove" Chart 3544;
anchor: 50°21.47' N, 125°41.95' W (NAD 83)

Noble Islets Light Chart 3549;
position: (W. pt of westerly Noble Islets):
50°49.30' N, 127°35.47' W (NAD 83)\

Noble Lagoon Chart 3710 (inset);
anchor (0.25 mile northwest Archer Islets):
52°31.53' N, 129°02.22' W (NAD 27)

Nodales Channel Chart 3543;
position north entrance:
50°26.50' N, 125°18.03' W (NAD 27)

Nootka Chart 3665;
resort: 49°37.43' N, 126°37.30' W (NAD 27)

Nootka Light Chart 3664;
position: (on summit of San Rafael Isl.):
49°35.57' N, 126°36.84' W (NAD 27)

Nordstrom Cove Chart 3686;
anchor west: 50°29.22' N, 127°55.61' W;
anchor east: 50°29.15' N, 127°55.40' W
(NAD 83)

Norman Morrison Bay Chart 3787;
entrance: 52°12.10' N, 128°11.60' W;
anchor: 52°11.47' N, 128°10.54' W (NAD 27)

North Arm Breakwater Light
Chart 3491;
position: (on breakwater):
49°15.27' N, 123°15.97' W (NAD 83)

North Arm Second Light Chart 3491;
position: (S. side of channel):
49°15.07' N, 123°16.00' W (NAD 83)

North Bentinck Arm Chart 3730;
entrance: 52°19.50' N, 126°18.50' W
(NAD 27)

"North Congreve Cove" Chart 3671;
anchor: 48°55.74' N, 125°01.50' W (NAD 27)

"North Cove," Helmcken Island,
Chart 3544;
anchor: 50°24.23' N, 125°52.56' W (NAD 83)

**North Cove: Cufra Inlet, Thetis
Island** Chart 3313, p. 15;
anchor (Cufra Inlet): 49°00.98' N,
123°41.23' W (NAD 83)

North Hanmer Light Chart 3717;
position: 54°03.83' N, 130°14.62' W
(NAD 83)

North Harbour Chart 3686;
anchor: 50°29.15' N, 128°02.70' W (NAD 83)

North Nomad ODAS Light Buoy
46184 Chart 3000;
position: 53°54.00' N, 138°52.00' W
(NAD 27)

North Pacific Cannery Museum
Chart 3717;
position: 54°11.65' N, 130°13.40' W (NAD 27)

North Rachael Island Light
Chart 3957;
position: (N. side of isl.):
54°12.75' N, 130°33.57' W (NAD 83)

North Reef Light Chart 3442;
position: (on reef): 48°54.85' N,
123°37.53' W (NAD 27)

"North Shortcut" Charts 3957, 3955;
south entrance: 54°19.63' N, 130°29.38' W;
north entrance: 54°20.57' N, 130°29.15' W
(NAD 83)

Northeast Bay Chart 3513;
position: 49°42.92' N, 124°22.86' W
(NAD 27)

Northeast Point Light Chart 3311;
position: (E. end of pt, Texada Isl.):
49°42.53' N, 124°21.38' W (NAD 83)

Northside Jetty 1 Light Chart 3490;
position: (off westerly extremity of Steveston
Jetty): 49°06.35' N, 123°18.11' W (NAD 83)

Northside Jetty 1A Light Chart 3490;
position: (on jetty): 49°06.53' N,
123°17.66' W (NAD 83)

Northside Jetty 3 Light Chart 3490;
position: (on jetty):
49°06.73' N, 123°17.18' W (NAD 83)

Northside Jetty 3A Light Chart 3490;
position: 49°06.92' N, 123°16.74' W
(NAD 83)

Northside Jetty 5 Light Chart 3490;
position: (on jetty):
49°07.18' N, 123°16.10' W (NAD 83)

Northside Jetty 5A Light Chart 3490;
position: 49°07.45' N, 123°14.46' W
(NAD 83)

Northside Jetty 7 Light Chart 3490;
position: (on jetty, E. of No. 5):
49°07.71' N, 123°14.83' W (NAD 83)

Northside Jetty 7A Light Chart 3490;
position: (close behind Steveston Jetty):
49°07.87' N, 123°14.42' W (NAD 83)

Northside Jetty 11 Light Chart 3490;
position: (on jetty):
49°07.93' N, 123°13.50' W (NAD 83)

Northside Jetty 13 Light Chart 3490;
position: (northside of jetty):
49°07.80' N, 123°12.76' W (NAD 83)

Northumberland Channel
Chart 3313, p. 20;
south entrance position:
49°08.45' N, 123°49.20' W (NAD 83)

Northwest Bay Chart 3459;
position: 49°18.07' N, 124°11.93' W (NAD 27)

"Northwest Hecate Island Cove"
Chart 3784;
position: 51°42.18' N, 128°03.22' W (NAD 27)

Northwest Rocks Light Chart 3747;
position: (on largest rock of group):
53°32.87' N, 130°38.03' W (NAD 27)

Norway Island Light Chart 3477;
position: (NW end of Norway Isl.):
48°58.90' N, 123°37.57' W (NAD 27)

Nose Point Light Chart 3478;
position: (extremity of pt):
48°50.57' N, 123°25.09' W (NAD 83)

Nowish Cove Chart 3734;
north entrance: 52°31.60′ N, 128°26.20′ W;
anchor: 52°31.33′ N, 128°25.64′ W (NAD 27)

Nuchatlitz Chart 3663;
Nuchatlitz light: 49°49.22′ N, 126°58.83′ W;
anchor: 49°48.55′ N, 126°57.54′ W (NAD 27)

Nuchatlitz Inlet Chart 3663
position: 49°46.50′ N, 126°56.90′ W
(NAD 27)

Nuchatlitz Light Chart 3663;
position: (NW extremity of unnamed isl.):
49°49.22′ N, 126°58.83′ W (NAD 27)

Nugent Sound Chart 3552;
west entrance: 51°05.52′ N, 127°27.95′ W
(NAD 27)

"Nugent Sound Cove" Chart 3552;
anchor island (69):
51°05.56′ N, 127°23.28′ W (NAD 27)

Numas Island Light Chart 3547;
position: (SE part of isl.):
50°45.92′ N, 127°04.33′ W (NAD 83)

Nuttal Bay Charts 3459, 3512;
position: 49°18.40′ N, 124°11.50′ W (NAD 27)

Nymphe Cove Chart 3539 (inset);
anchor: Nymphe Cove:
50°07.75′ N, 125°21.88′ W (NAD 83)

O'Brien Bay, "Rusty Cove"
Chart 3515 (inset);
anchor Rusty Cove:
50°51.36′ N, 126°32.87′ W (NAD 83)

O'Leary Islets Chart 3683;
"Bowler Islet": 50°06.13′ N, 127°39.87′ W
(NAD 27)

Oak Bay Chart 3313, p. 4; Chart 3424;
Mary Tod Island entrance light:
48°25.55′ N; 123°17.92′ W (NAD 83)

Oatswish Bay Chart 3962;
entrance: 52°54.50′ N, 128°07.60′ W;
beach: 52°55.70′ N, 128°07.80′ W (NAD 27)

Observatory Inlet Chart 3933;
entrance; 55°00.70′ N, 130°01.60′ W (NAD 27)

Observatory Inlet Light Chart 3933;
position: (on small isl.):
55°09.90′ N, 129°54.02′ W (NAD 27)

Obstruction Islet Chart 3552;
position: 51°08.56′ N, 127°06.82′ W (NAD 27)

Ocean Falls Chart 3781; floats:
52°21.14′ N, 127°41.76′ W (NAD 27)

Ochwe Bay Chart 3743;
position (Boulder Creek):
53°29.25′ N, 128°45.28′ W (NAD 27)

Octopus Islands Marine Park
Charts 3537, 3539;
anchor (west): 50°16.75′ N, 125°13.85′ W;
(east): 50°16.78′ N, 125°13.57′ W (NAD 27)

Octopus Point Light Chart 3478;
position: 48°47.96′ N, 123°33.68′ W
(NAD 83)

Odin Cove Chart 3787, 3720;
entrance: 52°12.80′ N, 128°10.70′ W;
anchor: 52°12.56′ N, 128°10.67′ W (NAD 27)

Odlum Island Light Chart 3784;
position: (W. side of pt):
51°41.62′ N, 128°07.12′ W (NAD 27)

Ogden Channel Charts 3773, 3927;
south entrance: 53°49.60′ N, 130°18.70′ W;
north entrance: 53°55.40′ N, 130°14.50′ W
(NAD 27)

Ogden Point Breakwater Light
Chart 3415;
position: (on outer end of breakwater):
48°24.82′ N, 123°23.55′ W (NAD 27)

Ohlsen Point Light Chart 3681;
position: (on the pt):
50°32.06' N, 127°34.19' W (NAD 83)

Okeover Inlet Charts 3559, 3312, 3538;
entrance: 50°02.00' N, 124°44.50' W (NAD 27)

Okeover Landing
Charts 3559, 3312, 3538;
public float: 49°59.49' N, 124°42.57' W
(NAD 27)

Okisollo Channel Chart 3537
west entrance: 50°17.20' N, 125°22.80' W
(NAD 27)

Old House Bay Chart 3512;
anchor 49°27.55' N, 124°15.50' W (NAD 27)

Oldfield Breakwater Light Chart 3958;
position: (on floating breakwater):
54°17.70' N, 130°21.25' W (NAD 83)

Oliver Cove Charts 3710 (inset), 3728;
entrance: 52°18.75' N, 128°21.18' W;
anchor: 52°18.69' N, 128°21.05' W (NAD 27)

**"One Foot Rock Cove,"
Bischof Islands** Chart 3808;
entrance (west): 52°34.37' N, 131°34.04' W;
anchor ("One Foot Rock"):
52°34.48' N, 131°33.40' W (NAD 27)

One Tree Islet Light Chart 3963;
position: (N. side of islet):
54°33.87' N, 130°26.34' W (NAD 83)

Oona River Chart 3773;
entrance: 53°56.44' N, 130°14.13' W
(NAD 27)

Open Bay Charts 3539, 3538, 3312;
entrance: 50°07.80' N, 125°12.20' W
(NAD 83)

Open Bight Chart 3934;
anchor: 51°22.17' N, 127°46.44' W (NAD 83)

Open Cove Chart 3545, 3564;
anchor: 50°32.26' N, 126°16.33' W (NAD 83)

Orford Bay Chart 3542 or 3312;
position: 50°35.60' N, 124°51.85' W
(NAD 83)

Ormidale Harbour Charts 3787, 3720;
entrance: 52°12.40' N, 128°09.00' W (NAD 27)

Ormiston Point Light Chart 3772;
position: (E. shore of Pitt Isl., Grenville
Channel): 53°34.25' N, 129°39.42' W
(NAD 27)

Osborn Bay, Crofton Chart 3475 (inset);
wharf light: 48°51.95' N, 123°38.20' W
(NAD 27)

Osborne Island Light Chart 3957;
position: 54°17.20' N, 130°51.19' W
(NAD 83)

Oscar Passage to Klemtu Chart 3734;
east entrance: 52°29.00' N, 128°16.00' W;
west entrance: 52°27.50' N, 128°25.00' W
(NAD 27)

Osment Inlet
Chart 3737 (inner basin uncharted);
entrance: 52°32.30' N, 128°43.50' W;
anchor (approximately):
52°31.88' N, 128°41.87' W (NAD 27)

Oswald Bay Chart 3724;
entrance: 53°01.03' N, 129°41.00' W (NAD 27)

Otter Bay, Hyashi Cove
Chart 3313, p. 8;
position: 48°47.85' N, 123°18.50' W
(NAD 83)

Otter Channel Chart 3742;
west entrance: 53°11.50' N, 129°35.00' W;
east entrance: 53°12.00' N, 129°29.50' W
(NAD 27)

Otter Cove Chart 3539;
anchor: 50°19.47' N, 125°27.10' W (NAD 83)

Otter Passage Chart 3742;
west entrance: 53°06.85' N, 129°46.60' W;
entrance: 53°09.40' N, 129°42.80' W (NAD 27)

Otter Passage Light Chart 3742;
position: (S. end of Man Isl.):
53°07.70' N, 129°46.33' W (NAD 27)

Ououkinsh Inlet Chart 3683
entrance: 50°07.30' N, 127°35.90' W (NAD 27)

"Outer Cove" Chart 3772;
south entrance: 53°22.39' N, 129°27.42' W;
anchor (outer cove):
53°22.65' N, 129° 27.05' W (NAD 27)

Outer Narrows, Slingsby Channel
Chart 3550;
position: 51°05.25' N, 127°37.85' W (NAD 83)

**Oval Bank West Cardinal Light Buoy
EOB** Chart 3927;
position: 53°55.60' N, 130°54.30' W (NAD 27)

Owen Bay Chart 3537;
public float: 50°18.65' N, 125°13.33' W
(NAD 27)

Owyacumish Bay Chart 3745;
entrance: 53°29.70' N, 128°22.00' W;
anchor (waterfall):
53°30.53' N, 128°22.03' W (NAD 27)

Oyster Bay Chart 3921;
anchor (north): 51°37.85' N, 127°41.34' W;
anchor (east): 51°37.55' N, 127°41.28' W
(NAD 83)

Oyster Bay Charts 3538, 3513;
anchor: 49°56.10' N, 125°10.70' W (NAD 27)

Oyster River Chart 3513;
entrance: 49°52.30' N, 125°06.45' W
(NAD 27)

Pachena Bay Chart 3602
entrance: 48°45.80' N, 125°09.00' W
(NAD 27)

Pachena Point Chart 3602;
light: 48°43.34' N, 125°05.77' W (NAD 27)

Pachena Point Light Chart 3602;
position: (on the pt):
48°43.34' N, 125°05.77' W (NAD 27)

Pacofi Bay Chart 3811(inset), 3894;
entrance: 52°50.50' N, 131°51.60' W;
anchor: 52°50.10' N, 131°52.65' W
(CHS states NAD unknown)

Palmer Anchorage Chart 3737;
position: 52°37.30' N, 128°45.50' W (NAD 27)

Pam Rock Light Chart 3526;
position: (on rock):
49°29.27' N, 123°17.90' W (NAD 83)

Pamphlet Cove ("Quiet Cove")
Chart 3679;
anchor: 50°31.11' N, 127°39.30' W (NAD 83)

**"Panther Point Cove," Wallace
Island** Chart 3313, p. 17;
anchor: 48°55.98' N, 123°32.15' W (NAD 83)

Panther Point Light Buoy U44
Chart 3442;
position: (W. of pt):
48°55.80' N, 123°32.43' W (NAD 27)

Paradise Passage Chart 3963;
south entrance: 54°37.85' N, 130°23.60' W;
north entrance: 54°38.90' N, 130°23.38' W
(NAD 83)

Parizeau Point Light Chart 3958;
position: (on dolphin):
54°17.29' N, 130°22.15' W (NAD 83)

Parker Harbour Charts 3559, 3312, 3538;
anchor: 50°03.18' N, 124°47.38' W (NAD 27)

Parry Passage Lght Chart 3895;
position: (eastern extremity of Lucy Isl.):
54°10.91' N, 132°58.29' W (NAD 27)

Parry Patch Sector Light Chart 3711;
position: 52°40.48' N, 128°31.93' W
(NAD 27)

Parson Island Light Chart 3546;
position: (on drying rock):
50°34.51' N, 126°41.00' W (NAD 83)

Parsons Anchorage Chart 3737;
anchor: 52° 31.00' N, 128° 44.30' W
(NAD 27)

Pasley Island Chart 3526;
position: 49°22.18' N, 123°27.23' W (NAD 27)

Passage Cove Charts 3710 (inset), 3728;
entrance: 52°18.48' N, 128°21.44' W
(NAD 27)

Patey Rock Light Chart 3441;
position: (on rock, entrance to Saanich Inlet):
48°42.05' N, 123°31.17' W (NAD 27)

Patricia Bay Breakwater Light
Chart 3441;
position: 48°39.15' N, 123°26.98' W
(NAD 27)

Patrician Cove Chart 3548;
anchor: 50°43.48' N, 127°25.06' W (NAD 83)

Patrol Islet Light Buoy U16
Chart 3476;
position: 48°41.97' N, 123°25.42' W
(NAD 27)

Patterson Inlet Chart 3721 (inset);
entrance: 53°26.00' N, 129°50.87' W;
anchor (north basin):
53° 27.42' N, 129°47.08' W (NAD 27)

Payne Bay Chart 3313 (inset), p. 9;
anchor: 48°52.36' N, 123°23.36' W (NAD 83)

Pearl Harbour Charts 3959, 3963;
entrance(Boat Passage):
54°30.16' N, 130°28.30' W;
anchor: 54°30.29' N, 130°26.83' W (NAD 83)

Pearse Canal Chart 3994; north
entrance: 55°02.20' N, 130°12.60' W (NAD 27)

Pearse Canal Island Light Chart 3963;
position: (S. end of isl.):
54°47.04' N, 130°36.60' W (NAD 83)

Pearse Passage Chart 3546;
south entrance: 50°34.50' N, 126°53.90' W
(NAD 83)

Pearson Point Light Chart 3920;
position: (on pt): 55°27.36' N, 129°30.08' W
(NAD 83)

Peet Bay Chart 3552;
anchor: 51°09.83' N, 127°04.23' W (NAD 27)

Peile Point Light Chart 3478;
position: (on the pt):
48°51.00' N, 123°24.23' W (NAD 83)

Pellow Islets Light Buoy U15
Chart 3476;
position: (E. of Portland Islands):
48°43.35' N, 123°20.97' W (NAD 27)

Pelly Island Light Chart 3415;
position: (S. extremity of isl.):
48°25.53' N, 123°22.98' W (NAD 27)

Pemberton Bay Chart 3723 (inset);
entrance: 52°56.20' N, 129°35.40' W (NAD 27)

Pender Canal Chart 3313 (inset), p. 12;
south entrance: 48°45.74' N, 123°15.45' W
(NAD 83)

Pender Harbour Light Chart 3535;
position: (on reef, westward of N. extreme of
Williams Isl.): 49°37.79' N, 124°03.57' W
(NAD 27)

Pendrell Sound Charts 3541, 3312;
entrance position:
50°12.40′ N, 124°45.00′ W (NAD 27)

Penn Harbour Chart 3737;
entrance: 52°58.05′ N, 128°58.25′ W;
anchor: 52°58.47′ N, 128°57.12′ W (NAD 27)

Penrose Bay Charts 3559, 3312, 3538;
anchor: 50°00.53′ N, 124°43.59′ W (NAD 27)

**Penrose Island and
Walbran Island** Chart 3934;
position (Rouse Reef):
51°29.10′ N, 127°46.67′ W (NAD 83)

Perceval Narrows Chart 3728;
south entrance: 52°19.85′ N, 128°22.55′ W;
north entrance: 52°20.17′ N, 128°22.47′ W
(NAD 27)

Percy Anchorage Chart 3313, p. 20 (inset);
south entrance: 49°08.45′ N, 123°49.20′ W
(NAD 83)

Perrin Anchorage Light Chart 3710;
position: (E. of Ivory Isl.):
52°16.48′ N, 128°23.37′ W (NAD 27)

Perry Bay Chart 3920;
entrance (north): 55°23.90′ N, 129°40.90′ W;
anchor (shelf): 55°22.77′ N, 129°40.90′ W
(NAD 83)

Peter Bay Chart 3787; northeast
entrance: 52°05.75′ N, 128°16.50′ W;
anchor: 52°04.50′ N, 128°16.95′ W (NAD 27)

Peter Cove Chart 3313, p. 11;
anchor: 48°44.32′ N, 123°14.02′ W (NAD 83)

Peters Narrows Chart 3772;
position: 53°22.62′ N, 129°27.38′ W (NAD 27)

Peterson Islet Light Chart 3544;
position: (N. end of islet):
50°23.38′ N, 125°54.80′ W (NAD 83)

Petley Point Float Chart 3515;
float: 50°55.72′ N, 126°13.15′ W (NAD 83)

Petrel Channel Narrows Chart 3746;
position: 53°42.60′ N, 130°09.00′ W
(NAD 27)

Petrel Channel North Chart 3747;
north entrance: 53°49.00′ N, 130°18.00′ W
(NAD 27)

Petrel Channel South Chart 3746;
south entrance: 53°33.60′ N, 130°02.00′ W
(NAD 27)

Petrel Rock Light and Bell Buoy D39
Chart 3958;
position: (S. side of rock):
54°13.67′ N, 130°24.87′ W (NAD 83)

"Petroglyph Cove" Chart 3682;
anchor: 50°01.5′ N 127°10.7′ W (NAD 27)

Philip Inlet Chart 3934;
entrance: 51°33.33′ N, 127°47.20′ W;
anchor: 51°34.18′ N, 127°45.35′ W (NAD 83)

Philip Narrows Chart 3737;
fairway position:
52°42.68′ N, 129°00.24′ W (NAD 27)

Phillimore Point Light Chart 3473;
position: (on extremity of pt):
48°52.30′ N, 123°23.43′ W (NAD 27)

Phillips Arm Chart 3543;
position: 50°31.80′ N, 125°23.87′ W
(NAD 27)

"Phoenix Creek Cove
Charts 3761, 3927;
west entrance: 53°53.65′ N, 130°31.40′ W;
anchor 53°53.53′ N, 130°30.77′ W (NAD 27)

"Pictograph Passage" Chart 3787
position: 52°03.90′ N, 128°14.75′ W (NAD 27)

Pierce Bay Chart 3934;
anchor (West Cove):
51°31.96' N, 127°46.00' W;
anchor (East Cove):
51°32.23' N, 127°45.00' W (NAD 83)

Piers Island Light Buoy U12
Chart 3476;
position: 48°42.25' N, 123°23.90' W (NAD 27)

Pill Point Light Chart 3671;
position: 48°57.95' N, 125°04.83' W
(NAD 27)

Pine Island Light Chart 3549;
position: (SW pt of isl.):
50°58.55' N, 127°43.59' W (NAD 83)

Pinkerton Islands Chart: 3670;
anchor: 48°57.56' N, 125°16.66' W (NAD 83)

Pipestem Inlet Chart 3670
entrance: 49°01.20' N, 125°18.85' W (NAD 27)

**Pirates Cove Marine Park, De
Courcy Island** Chart 3313, p. 18 (inset);
anchor: 49°05.87' N, 123°43.83' W (NAD 83)

Pitt Island (Grenville Channel) Light
Chart 3772;
position: (on shore):
53°42.28' N, 129°48.97' W (NAD 27)

Pitt Point Light Buoy D3 Chart 3773;
position: 53°53.43' N, 130°06.05' W
(NAD 27)

Pitt Point Light Chart 3773;
position: (on pt, SE of Calvert Pt):
53°52.97' N, 130°06.57' W (NAD 27)

Plover Point Light Chart 3742;
position: 53°16.80' N, 129°18.20' W
(NAD 27)

Plumper Bay Chart 3539 (inset);
anchor: 50°09.48' N, 125°20.36' W (NAD 83)

**Plumper Cove and Plumper Cove
Marine Park** Chart 3535 (inset);
anchor: 49°24.08' N, 123°28.50' W (NAD 83)

Plumper Harbour Chart 3664;
anchor: 49°41.53' N, 126°37.80' W (NAD 27)

Pocahontas Bay Chart 3513;
position: 49°43.56' N, 124°25.72' W
(NAD 27)

Pocahontas Point Light Chart 3668;
position: (on pt): 48°59.00' N, 124°55.07' W
(NAD 83)

Poett Nook Chart 3671;
anchor: 48°52.70' N, 125°03.01' W (NAD 27)

Point Atkinson Chart 3481;
light: 49°19.83' N, 123°15.80' W (NAD 27)

Point Atkinson Light Chart 3481;
position: (N. point of entrance to Burrard
Inlet): 49°19.83' N, 123°15.80' W (NAD 27)

Point Cowan Light Chart 3481;
position: (on the pt):
49°20.14' N, 123°21.58' W (NAD 27)

Point Cumming Light Chart 3740;
position: (on the pt, SW end of Gribbell Isl.):
53°18.72' N, 129°07.25' W (NAD 27)

Point Fairfax Light Chart 3441;
position: (S. extremity of pt):
48°41.96' N, 123°17.84' W (NAD 27)

Point Grey Light and Bell Buoy Q62
Chart 3481;
position: (NW of pt):
49°17.35' N, 123°15.90' W (NAD 27)

Point Roberts to Point Grey
Chart 3463;
Point Roberts marina entrance:
49°58.35' N, 123°03.80' W (NAD 27)

Point Upwood Light Chart 3512;
position: (southeastern extreme of Texada
Island): 49°29.37' N, 124°08.38' W (NAD 27)

Pointer Island Light Chart 3785;
position: (SE end of isl., S. of E. entrance to
Lama Passage): 02°03.65' N, 127°56.80' W
(NAD 27)

Pointer Rocks Light Chart 3959;
position: (on southernmost rock):
54°36.29' N, 130°32.25' W (NAD 83)

Poison Cove Chart 3962;
entrance: 52°54.60' N, 128°02.30' W
(NAD 27)

Polly Point Light Chart 3668;
position: (on pt): 49°12.96' N, 124°49.02' W
(NAD 83)

Popham Island Light Chart 3526;
position: 49°21.75' N, 123°29.43' W
(NAD 83)

Porcher Inlet Charts 3761, 3927;
entrance: 53°52.70' N, 130°30.70' W
(NAD 27)

Porlier Pass ("Cowichan Gap")
Chart 3313, p. 16;
Virago Rock light: 49°00.78' N,
123°35.49' W (NAD 83)

Porlier Pass Light and Bell Buoy U41
Chart 3473;
position: (off E. entrance to pass):
49°01.56' N, 123°34.88' W (NAD 27)

Porlier Pass Range Light Chart 3473;
position: (on Race Pt, Galiano Isl.):
49°00.79' N, 123°35.07' W (NAD 27)

Porpoise Bay Chart 3312;
public float: 49°29.03' N, 123°45.40' W
(NAD 27)

Porpoise Channel Charts 3955, 3958;
entrance range (0.14 mile NW Buoy):
54°11.53' N, 130°20.50' W;
(course 048° M); middle range (0.09 mile
south of buoy "D35"):
54°11.88' N, 130°30.57' W (course 015° M);
(NAD 27)

Porpoise Channel East Light
Chart 3955;
position: (on drying rock, E. side of channel):
54°12.70' N, 130°17.44' W (NAD 27)

Porpoise Channel Entrance Light
Chart 3955;
position: (on drying rock, S. side of channel):
54°12.07' N, 130°18.18' W (NAD 27)

Porpoise Channel West Light
Chart 3955;
position: (NW side of channel):
54°12.47' N, 130°17.86' W (NAD 27)

Porpoise Harbour Charts 3955, 3958;
outer entrance (073°):
54°11.53' N, 130°20.50' W;
inner entrance (039.5°):
54°11.88' N, 130°18.58' W;
anchor (north end):
54°14.64' N, 130°18.67' W (NAD 27)

**Porpoise Harbour Entrance
Range Light** Chart 3955;
position: (on Flora Bank, W. side of Lelu Isl.):
54°12.00' N, 130°17.94' W (NAD 27)

Porpoise Harbour Light Chart 3955;
position: (N. side of entrance to hbr):
54°11.92' N, 130°18.93' W (NAD 27)

Port Albion Chart 3646
position: 48°57.00' N, 125°32.60' W (NAD 27)

Port Alexander Chart 3549;
anchor: 50°51.40' N, 127°39.90' W (NAD 83)

Port Alice Chart 3681;
position: 50°23.00' N, 127°27.00' W (NAD 83)

Port Browning Chart 3313, p. 11;
anchor: 48°46.57' N, 123°16.10' W (NAD 83)

Port Canaveral, Canaveral Passage
Charts 3753 (inset), 3746;
anchor (nook west end of Canaveral
Passage): 53°34.17' N, 130°08.55' W;
anchor (Dixon Island nook):
53°35.08' N, 130°10.16' W (NAD 27)

Port Désiré Chart 3646;
anchor: 48°49.83' N, 125°07.44' W (NAD 83)

Port Eliza Chart 3663
entrance: 49°51.45' N, 126°58.60' W (NAD 27)

Port Elizabeth, Duck Cove Chart 3545;
anchor: 50°39.87' N, 126°30.58' W (NAD 83)

Port Graves Chart 3526;
anchor: 49°28.35' N, 123°21.30' W (NAD 27)

Port Hardy Aeronautical Light
Chart 3548;
position: 50°41.07' N, 127°22.45' W (NAD 83)

Port Hardy Boat Basin North Light
Chart 3548;
position: (seaward end, N. rock breakwater):
50°42.86' N, 127°29.32' W (NAD 83)

Port Hardy Boat Basin South Light
Chart 3548;
position: (seaward end, S. rock breakwater):
50°42.79' N, 127°29.30' W (NAD 83)

Port Hardy Chart 3548 (inset);
inner light (SE of wharf):
50°43.12' N, 127°28.88' W
(shoal, stay east!) (NAD 83)

Port Harvey, Mist Islets
Chart 3564,3545;
anchor site north of Range Island:
50°33.95' N, 126°16.25' W (NAD 83)

Port John Chart 3785;
entrance: 52°07.20' N, 127°51.20' W;
anchor: 52°07.32' N, 127°50.40' W (NAD 27)

Port Langford Chart 3663;
anchor: 49°48.76' N, 126°56.67' W (NAD 27)

**Port Mann Training Dyke Lower
Light** Chart 3490;
position: (on channel end of dyke):
49°12.91' N, 122°53.02' W (NAD 83)

**Port Mann Training Dyke Upper
Light** Chart 3490;
position: (on channel end of dyke):
49°13.02' N, 122°52.68' W (NAD 83)

Port McNeill Breakwater Light
Chart 3546;
position: 50°35.57' N, 127°05.30' W
(NAD 83)

Port McNeill Chart 3546 (inset);
breakwater light: 50°35.58' N, 127°05.20' W
(NAD 83)

Port Moody Chart 3495;
anchor: 49°17.35' N, 122°51.25' W
(NAD 83)

Port Moody Light Chart 3495;
position: (entrance to dredged area):
49°17.08' N, 122°51.11' W (NAD 83)

Port Moody Range Light Chart 3495;
position: (on the N. shore):
49°17.59' N, 122°51.27' W (NAD 83)

Port Neville Charts 3564, 3545;
public float: 50°29.57' N, 126°05.26' W
(NAD 83)

**Port San Juan Light and Whistle
Buoy YK** Chart 3647;
position: (at entrance):
48°32.08' N, 124°29.02' W (NAD 27)

Port Simpson (Lax Kw' Alaams)
Charts 3963 (inset), 3959;
breakwater light: 54°33.71' N, 130°25.71' W
(NAD 83)

Port Simpson Breakwater Light
Chart 3963;
position: 54°33.72' N, 130°25.71' W
(NAD 83)

Port Stephens Chart 3742;
entrance: 53°19.30' N, 129°42.50' W (NAD 27)

Port Washington Light Chart 3442;
position: (on Boat Isl.):
48°48.70' N, 123°19.33' W (NAD 27)

Portage Cove Charts 3312, p.9, 3538;
anchor: 50°05.13' N, 124° 43.45' W
(NAD unknown)

Porteau Cove Marine Park Chart 3526;
dive wrecks: 49°33.67' N, 123°14.05' W
(NAD 27)

Portland Canal Chart 3933;
entrance: 54°58.80' N, 130°07.80' W (NAD 27)

Portland Inlet Chart 3994;
entrance: 54°41.00' N, 130°27.50' W (NAD 27)

Portlock Point Light Chart 3442;
position: (on the pt):
48°49.68' N, 123°21.03' W (NAD 27)

Potts Bay (Midsummer Island)
Chart 3546;
position: 50°39.00' N, 126°37.30' W
(NAD 83)

"Potts Island Hideaway" Chart 3787;
entrance: 52°08.50' N, 128°19.90' W;
anchor: 52°08.65' N, 128°20.43' W (NAD 27)

Potts Lagoon Chart 3545;
lagoon: 50°33.55' N, 126°27.12' W (NAD 83)

"Potts Lagoon East Basin" Chart 3545;
east basin: 50°33.95' N, 126°27.03' W
(NAD 83)

Powell River
Charts 3311, sheet 5; 3536 (inset),
Westview south basin breakwater light:
49°50.03' N, 124°31.79' W (NAD 83)

**Powell River Floating Breakwater
Entrance North Light** Chart 3536;
position: (N. side of entrance):
49°51.87' N, 124°33.47' W (NAD 27)

**Powell River Floating Breakwater
Entrance South Light** Chart 3536;
position: (S. side of entrance):
49°51.78' N, 124°33.38' W (NAD 27)

"Power Wash Waterfall" Chart 3552,
position: 51°08.35' N, 127°18.40' W
(NAD 27)

Preedy Harbour Chart 3313, p. 16 (inset);
public float: 48°58.79' N, 123°40.70' W;
anchor: 48°58.76' N, 123°40.87' W (NAD 83)

Preedy Harbour Light Chart 3477;
position: (off W end of Hudson Isl.):
48°58.09' N, 123°40.98' W (NAD 27)

Price Cove Chart 3745;
position: 53°16.05' N, 127°56.80' W
(NAD 27)

"Price Inlet" Chart 3728 (uncharted);
entrance (uncharted inlet):
52°29.10' N, 128°40.30' W (NAD 27)

Prideaux Haven Chart 3555;
anchor: 50°08.67' N, 124°40.71' W (NAD 27)

Prideaux Island Light Chart 3670;
position: 48°56.58' N, 125°16.15' W
(NAD 83)

Priestland Cove, Halfmoon Bay
Chart 3535 (Welcome Passage inset);
Priestland Cove public wharf:
49°30.62' N, 123°54.67' W (NAD 27)

Prince Rupert Aeronautical Light
Chart 3958;
position: 54°17.40' N, 130°26.47' W
(NAD 83)

Prince Rupert Harbour Chart 3958;
south entrance (midchannel between Barret
Rock and Buoy "D47"):
54°14.74' N, 130°20.86' W (NAD 27)

Princesa Channel Light Chart 3665;
position: (edge of reef at E. entrance to
channel): 49°43.42' N, 126°37.50' W
(NAD 27)

Princess Bay ("Tortoise Bay")
Chart 3313 (inset), p. 9;
anchor 48°43.10' N, 123°22.20' W (NAD 83)

"Princess Bay" Charts 3313 (p. 17); 3442;
anchor: 48°56.56' N, 123°33.33' W (NAD 83)

Princess Louisa Inlet Chart 3312
entrance: 50°09.60' N, 123°51.00' W
(NAD unknown)

**Princess Margaret Marine Park,
Portland Island** Chart 3313 (inset), p. 9
See Royal Cove.

Princess Royal Channel
Charts 3738, 3739;
south entrance Graham Reach:
52°53.75' N, 128°30.60' W (NAD 27)

Principe Channel Chart 3742;
south entrance: 53°15.00' N, 129°42.00' W
(NAD 27)

Principe Channel Light Chart 3741;
position: (on shore):
53°23.60' N, 129°54.30' W (NAD 27)

Principe Channel, North Entrance
Chart 3747;
Principe Channel (north entrance):
53°39.50' N, 130° 26.00' W (NAD 27)
(See Chapter 11 for Browning Entrance and
Larsen Harbour.)

Prospect Point Light Chart 3493;
position: (under bluff at pt):
49°18.85' N, 123°08.41' W (NAD 83)

Protection Cove Chart 3550;
position: 51°14.06' N, 127°47.00' W
(NAD 83)

Providence Passage Chart 3546;
center of passage:
50°39.00' N, 126°41.10' W (NAD 83)

Pruth Bay Chart 3784, 3727;
anchor: 51°39.27' N, 128°07.52' W
(NAD 27)

Pryce Channel Light Chart 3541;
position: (Redonda Islands):
50°18.42' N, 124°49.72' W (NAD 83)

Pulteney Point Chart 3546;
light: 50°37.85' N, 127°09.20' W (NAD 83)

Pulteney Point Light Chart 3546;
position: (on pt, SW end of Malcolm Isl.):
50°37.83' N, 127°09.29' W (NAD 83)

Pulton Bay Chart 3537;
position: 50°18.27' N, 125°16.27' W
(NAD 27)

Purcell Rock and Ironbound Islet
Chart 3730;
position: 52°39.40' N, 127°01.00' W
(NAD 27)

Purfleet Point Range Light Chart 3490;
position: (W. extremity of Annacis Island):
49°09.56' N, 123°58.80' W (NAD 83)

Qlawdzeet ("Squatterie")
Chart 3909 (inset);
entrance: 54°13.00' N, 130°46.50' W;
anchor: 54°12.47' N, 130°45.87' W (NAD 83)

Quait Bay ("Calm Creek") Chart 3649;
anchor: 49°16.57' N, 125°51.05' W (NAD 27)

Quarantine Cove Charts 3641, 3440;
William Head light: 48°20.58' N, 123°31.56' W
(NAD 27)

Quarry Bay, Nelson Island
Chart 3311 or 3312;
entrance: 49°39.40' N, 124°08.10' W;
anchor: 49°39.57' N, 124°07.21' W (NAD 83)

Quarry Point Light Chart 3738;
position: (on pt, E. side of Princess Royal Isl.):
52°54.02' N, 128°31.25' W (NAD 27)

Quartcha Bay Chart 3940;
entrance: 52°30.10' N, 127°50.60' W;
anchor: 52°30.74' N, 127°50.18' W (NAD 83)

Quartz Bay Charts 3538, 3312
anchor: 50°09.37' N, 125°00.15' W (NAD 27)

Quascilla Bay Chart 3931;
entrance: 51°17.65' N, 127°22.20' W (NAD 83)

Quathiaski Cove Chart 3540;
public float: 50°02.57' N, 125°13.02' W;
anchor: 50°03.10' N, 125°13.46' W (NAD 83)

Quathiaski Cove Light Chart 3540;
position: (SE of Grouse Isl.):
50°02.71' N, 125°13.29' W (NAD 83)

Quatsino Light Chart 3686;
position: (SE end of Kains Isl.):
50°26.47' N, 128°01.95' W (NAD 83)

Quatsino Narrows Chart 3681
south entrance: 50°32.02' N, 127°34.60' W
north entrance: 50°33.48' N, 127°33.37' W
(NAD 83)

Quatsino Sound Chart 3681
position: 50°32.00' N, 127°36.00' W (NAD 83)

Queen Charlotte City, Bearskin Bay
Charts 3890 (inset), 3894;
entrance: 53°14.60' N, 132°02.00' W;
Queen Charlotte breakwater:
53°15.16' N, 132°04.35' W;
anchor: 53°15.10' N, 132°04.60' W (NAD 27)

Queen Cove Chart 3663;
anchor: 49°52.72' N, 126°58.90' W (NAD 27)

Queens Sound Chart 3786;
south entrance: 51°50.00' N, 128°22.00' W;
entrance (Golby Passage):
52°01.50' N, 128°26.00' W (NAD 27)

"Quigley Creek Cove" Chart 3737;
entrance position: 52°39.44' N, 128°45.20' W;
anchor: 52°39.37' N, 128°44.66' W (NAD 27)

"Quottoon Narrows Cove"
Chart 3963;
entrance (Quottoon Inlet):
54°24.75' N, 130°06.45' W;
anchor: 54°28.20' N, 130°04.18' W (NAD 83)

Race Passage East Light Chart 3544;
position: (on rock):
50°22.86' N, 125°48.80' W (NAD 83)

Race Point Chart 3539;
light: 50°06.81' N, 125°19.41' W (NAD 83)

Race Point Light Chart 3539;
position: (E. extremity of pt):
50°06.81' N, 125°19.41' W (NAD 83)

Race Rocks Charts 3641, 3440;
Great Race Rock light:
48°17.89' N, 123°31.80' W (NAD 27)

**Race Rocks East Cautionary Light
Buoy VG** Chart 3461;
position: 48°16.08' N, 123°27.75' W
(NAD 27)

Race Rocks Light Chart 3461;
position: (Great Race Rock):
48°17.89' N, 123°31.80' W (NAD 27)

**Race Rocks South Cautionary Light
Buoy VF** Chart 3461;
position: 48°14.08' N, 123°31.90' W
(NAD 27)

Racey Inlet Chart 3737;
entrance: 52°53.12' N, 129°06.64' W (NAD 27)

**Rachael Islands South Cardinal Light
and Whistle Buoy DSO** Chart 3957;
position: 54°11.20' N, 130°33.50' W
(NAD 83)

Rae Basin Chart 3640; inner basin
anchor: 49°28.29' N, 126°24.17' W (NAD 27)

Raft Cove Chart 3624;
anchor north side: 50°35.30' N, 128°14.90' W
(NAD 27)

"Rainy Bay Cove" Charts 3670, 3668;
anchor: 48°58.95' N, 125°02.02' W (NAD 83)

Rait Narrows Chart 3787;
south entrance: 52°12.45' N, 128°17.30' W;
north entrance: 52°13.10' N, 128°16.35' W
(NAD 27)

"Rait Narrows, South Cove"
Chart 3787;
entrance: 52°12.38' N, 128°17.33' W;
anchor (central cove):
52°12.02' N, 128° 16.85' W;
anchor (small nook):
52°12.00' N, 128°16.80' W (NAD 27)

Ramsay Arm Chart 3541;
entrance 50°20.7' N, 124°58.80' W (NAD 27)

"Ramsay Passage Cove" Chart 3808;
entrance: 52°34.52' N, 131°24.05' W;
anchor: 52°34.35' N, 131°23.76' W (NAD 27)

Ramsbotham Islands Chart 3737;
light: 52°41.92' N, 129°02.10' W (NAD 27)

Ramsbotham Islands Light Chart 3737;
position: 52°41.92' N, 129°02.10' W
(NAD 27)

Ramsden Point Light Chart 3920;
position: (extremity of pt, Portland Inlet):
54°59.01' N, 130°06.18' W (NAD 83)

Rankin Cove Chart 3649;
entrance: 49°10.5' N, 125°42.3' W (NAD 27)

Rattenbury Point Charts 3781;
anchor: 52°14.70' N, 127°45.12' W (NAD 27)

Raven Cove Chart 3720;
anchor: 52°14.93' N, 128°08.97' W (NAD 27)

Ray Point Range Light Chart 3564;
position: (on the pt): 50°34.82' N,
126°12.02' W (NAD 83)

"Raymond Passage Cove" Chart 3787;
anchor: 52°07.11' N, 128°17.06' W
(NAD 27)

Read Island Charts 3539, 3312;
Read Island Public float: 50.11.08' N,
125°05.28' W (NAD 83)

Reba Point Cove Chart 3787;
entrance: 52°09.25' N, 128°20.30' W;
anchor: 52°09.04' N, 128°20.35' W (NAD 27)

Rebecca Rock Light Chart 3311;
position: (on summit of rock): 49°48.83' N,
124°39.50' W (NAD 83)

Rebecca Spit Marine Park Chart 3538;
anchor: 50°06.28' N, 125°11.52' W (NAD 27)

Reception Point Light Chart 3535;
position: (Welcome Passage):
49°28.27' N, 123°53.20' W (NAD 27)

Redcliffe Point Light Chart 3739;
position: (on the pt, Graham Reach):
53°09.42' N, 128°38.15' W (NAD 27)

Redonda Bay Chart 3555;
position: 50°15.62' N, 124°57.41' W (NAD 27)

Reed Point Light Chart 3495;
position: (off S. shore):
49°17.53' N, 122°52.42' W (NAD 83)

Reef Island Chart 3933;
anchor (south cove):
55°04.76' N, 130°12.53' W;
anchor (west Reef Island):
55°04.93' N, 130°12.32' W (NAD 27)

Reef Point Sector Light Chart 3711;
position: 52°37.11' N, 128°30.82' W (NAD 27)

Refuge Bay Charts 3909 (inset), 3956;
anchor: 54°03.24' N, 130°32.44' W (NAD 83)

Refuge Cove Chart 3555;
public float: 50°07.43' N, 124°50.30' W
(NAD 27)

Refuge Cove Light Chart 3555;
position: (SW tip of W. Redonda Isl.):
50°06.96' N, 124°50.89' W (NAD 27)

**Refuge Island Boat Harbour Sector
Light** Chart 3892;
position: (at Mia-Kwun Indian village):
54°06.78' N, 132°18.65' W (NAD 27)

Refuge Island Chart 3670;
anchor: 49°01.64' N, 125°18.74' W (NAD 83)

Regatta Rocks Light Chart 3787;
position: 52°13.13' N, 128°08.42' W (NAD 27)

Reid Passage Charts 3710 (inset), 3728;
south entrance: 52°15.90' N, 128°23.28' W;
north entrance: 52°19.45' N, 128°20.90' W;
Carne Rock light: 52°18.12' N, 128°21.80' W
(NAD 27)

Reid Passage Light Chart 3710;
position: (Reid Passage):
52°18.12' N, 128°21.80' W (NAD 27)

"Remotesville Cove" Chart 3921;
anchor: 51°37.54' N, 127°42.94' W (NAD 83)

Repulse Point Light Buoy P42
Chart 3527;
position: (SW of pt): 49°28.78' N,
124°44.83' W (NAD 27)

Rescue Bay Charts 3711 (inset), 3734;
entrance: 52°31.23' N, 128°17.19' W;
anchor: 52°30.92' N, 128°17.17' W (NAD 27)

Resolution Cove Chart 3664;
anchor: 49°36.42' N, 126°31.83' W (NAD 27)

Restoration Bay Chart 3729;
entrance: 52°01.00' N, 127°38.90' W;
anchor (southeast of yellow triangle):
52°01.50' N, 127°38.00' W (NAD 27)

Retreat Cove Chart 3313, p. 17;
anchor: 48°56.39' N, 123°30.10' W (NAD 83)

Return Channel Charts 3940, 3720;
east entrance: 52°18.20' N, 127°58.20' W;
west entrance: 52°16.90' N, 128°12.60' W
(NAD 83)

Richard Rock Light Chart 3670;
position: 48°59.33' N, 125°19.61' W (NAD 83)

Richards Point Light Chart 3933;
position: (extremity of pt Observatory Inlet):
55°17.30' N, 129°48.82' W (NAD 27)

Richardson Bay Chart 3313, p. 8;
anchor: 48°49.64' N, 123°21.30' W (NAD 83)

Richardson Cove Chart 3512;
anchor: 49°27.60' N, 124°16.40' W (NAD 27)

Richmond Bay Chart 3547 (inset);
anchor: 50°53.04' N, 126°58.74' W (NAD 83)

Rideout Islets Chart 3649;
position: 49°09' N, 125°43' W (NAD 27)

Ridley Island Light and Bell Buoy D27 Chart 3955;
position: 54°11.95' N, 130°21.33' W (NAD 27)

Riley Cove Chart 3648;
anchor: 49°23.45' N, 126°13.37' W (NAD 27)

Ripley Bay Chart 3940;
entrance: 52°24.70' N, 127°53.30' W;
anchor: 52°25.38' N, 127°53.30' W (NAD 83)

Ripple Point Light Chart 3543;
position: (on the pt) 50°22.08' N, 125°34.70' W (NAD 83)

Ripple Rock Charts 3539, 3540;
position: 50°07.88' N, 125°21.26' W (NAD 83)

Ritchie Bay Chart 3649; south bay
anchor: 49°13.63' N, 125°53.98' W;
north nook anchor: 49°14.11' N, 125°53.64' W (NAD 27)

Rivers Inlet Chart 3932;
float: 51°41.10' N, 127°15.75' W (NAD 83)

Rivers Inlet Chart 3934;
west entrance: 51°25.00' N, 127°46.80' W (NAD 83)

Rix Island Light Chart 3743;
position: (northeasterly end of Rix Isl.):
53°31.55' N, 128°44.15' W (NAD 27)

Roaringhole Rapids (Nepah Lagoon) Chart 3547;
position: 50°57.23' N, 126°48.70' W (NAD 83)

Robbers Passage Chart inset 3668;
PAYC float: 48°53.58' N, 125°07.02' W (NAD 83)

Roberts Bank Cautionary Light Buoy TA Chart 3463;
position: (W. of bank):
49°04.43' N, 123°22.77' W (NAD 27)

Roberts Bank Cautionary Light Buoy TB Chart 3499;
position: 49°01.47' N, 123°08.55' W (NAD 27)

Roberts Bank Entrance Range Light Chart 3499;
position: 49°01.60' N, 123°08.42' W (NAD 27)

Roberts Bank Light Buoy T1 Chart 3499;
position: 49°00.57' N, 123°09.59' W (NAD 27)

Roberts Bank Light Buoy T2 Chart 3499;
position: 49°00.50' N, 123°09.14' W (NAD 27)

Roberts Bank Light Buoy T4 Chart 3499;
position: 49°00.70' N, 123°08.84' W (NAD 27)

Roberts Bank Light Buoy T6 Chart 3499;
position: 49°00.89' N, 123°08.56' W (NAD 27)

Roberts Bank Light Buoy T8 Chart 3499;
position: 49°01.22' N, 123°08.51' W (NAD 27)

Roberts Bank Light Chart 3490;
position: (edge of bank):
49°05.27' N, 123°18.54' W (NAD 83)

Roberts Bank Limit Range Light Chart 3499;
position: 49°01.47' N, 123°09,12' W (NAD 27)

Roberts Bay Chart 3313, p. 7;
anchor: 48°39.88' N, 123°23.90' W (NAD 83)

"Robertson Cove" Charts 3538, 3312;
position: 50°11.74' N, 124°58.43' W (NAD 27)

Robertson Point Sector Light
Chart 3955;
position: (on bank, W. of pt):
54°19.40' N, 130°24.47' W (NAD 27)

Robson Bight Chart 3545;
position: 50°29.30' N, 126°35.00' W (NAD 83)

Robson Cove Chart 3679;
entrance: 50°30.20' N, 127°52.00' W
(NAD 83)

Roche Point Light Chart 3495;
position: (S. extremity of pt): 49°18.03' N,
122°57.36' W (NAD 83)

Rock Inlet Chart 3785 (inset);
entrance: 51°51.80' N, 127°52.00' W
anchor: 51°52.42' N, 127°51.28' W (NAD 27)

Rock Point Light Chart 3543;
position: (on the pt):
50°20.47' N, 125°30.02' W (NAD 83)

Rockfish Harbour Chart 3811 (inset);
entrance: 52°52.93' N, 131°46.50' W;
anchor: 52°52.95' N, 131°48.98' W
(CHS states NAD unknown)

Roderick Cove Chart 3734;
entrance: 52°40.60' N, 128°21.45' W
(NAD 27)

Roffey Island Eveleigh Anchorage
Charts 3559, 3312;
anchor: 50°08.44' N, 124°41.63' W (NAD 27)

Rogers Reef Light Chart 3475;
position: (E. entrance, Gabriola Pass):
49°07.88' N, 123°41.38' W (NAD 27)

Rolling Roadstead Chart 3663;
anchor: 49°50.86' N, 127°02.34' W (NAD 27)

Rolston Island Light Chart 3651;
position: 50°01.29' N, 127°21.87' W
(NAD 83)

Rooney Point Light Chart 3895;
position: (off pt): 54°01.03' N, 132°10.00' W
(NAD 27)

Roquefeuil Bay Chart 3671;
anchor: 48°51.18' N, 125°06.76' W (NAD 27)

"Rosa Harbour" Chart 3663;
anchor: 49°49.71' N, 126°58.56' W (NAD 27)

**Roscoe Bay Marine Park, West
Redonda Island** Charts 3538, 3312 (inset);
anchor: 50°09.63' N, 124°45.94' W (NAD 27)

Roscoe Creek Chart 3940;
position: 52°28.14' N, 127°44.45' W (NAD 83)

Roscoe Inlet Chart 3940;
entrance: 52°19.40' N, 127°56.30' W (NAD 83)

Roscoe Narrows Chart 3940;
position: 52°26.85' N, 127°55.70' W (NAD 83)

Rose Harbour Charts 3855, 3825;
entrance: 52°09.40' N, 131°05.00' W;
outside mooring buoy:
52°09.12' N, 131°05.06' W (NAD 27)

Rose Point Light and Bell Buoy C26
Chart 3802;
position: (N. of pt):
54°13.00' N, 131°39.10' W (NAD 27)

**Rose Spit Light and Whistle Buoy
C25** Chart 3802;
position: (northeastward of Overfall Shoal):
54°14.90' N, 131°30.70' W (NAD 27)

Rosedale Rock Light Buoy V15
Chart 3461;
position: (SE of rock):
48°17.45' N, 123°31.48' W (NAD 27)

Rosenfeld Rock Light Buoy U59
Chart 3462;
position: (N. extremity of reef):
48°48.20' N, 123°01.57' W (NAD 27)

"Ross Passage Cove" Chart 3648;
anchor: 49°20.06' N, 126°02.90' W (NAD 27)

Rough Bay Chart 3546;
Rough Bay breakwater light:
50°38.50' N, 127°01.90' W (NAD 83)

Round Island Light Chart 3548;
position: 50°43.58' N, 127°21.83' W (NAD 83)

Rouse Bay Chart 3512;
anchor: 49°28.36' N, 124°11.01' W (NAD 27)

Rowley Bay Chart 3552 (inset);
anchor: 51°07.65' N, 127°36.55' W (NAD 27)

Royal Cove Chart 3313, p.9;
anchor: Royal Cove (north end):
48°44.75' N, 123°22.24' W (NAD 83)

Royston Chart 3527;
anchor: 49°39.20' N, 124°56.65' W (NAD 27)

"Rubby Dub Cove" Chart 3679;
anchor: 50°30.05' N, 128°50.49' W (NAD 83)

Rudolf Bay Charts 3728, 3726;
entrance: 52°24.45' N, 128°45.10' W;
anchor: 52° 24.95' N, 128°45.25' W
(NAD 27)

Rugged Island Light Chart 3680;
position: (N pt of N isl.): 50°19.17' N,
127°54.97' W (NAD 27)

Rugged Point Chart 3682;
anchor: 49°58.20' N, 127°14.60' W (NAD 27)

Rugged Point Light Chart 3682;
position: 49°58.23' N, 127°15.03' W
(NAD 27)

Rumble Beach Charts 3681, 3679;
wharf: 50°25.42' N, 127°29.13' W (NAD 83)

Rupert Inlet Chart 3679
entrance: 50°34.10' N, 127°32.40' W (NAD 83)

"Rupert Island Passage" Chart 3784;
south entrance: 51°48.84' N, 128°08.07' W;
north entrance: 51°49.95' N, 128°07.00' W
(NAD 27)

Rushbrook Passage Chart 3963;
entrance (south): 54°36.08' N, 130°26.62' W;
entrance (north): 54°36.30' N, 130°26.95' W
(NAD 83)

Rushbrooke Floats Chart 3958;
north end breakwater:
54°19.59' N, 130°18.27' W (NAD 27)

**Rushton Island Light and
Bell Buoy D72** Chart 3957;
position: 54°16.03' N, 130°48.45' W (NAD 83)

Russell Arm Chart 3958;
entrance: 54°19.27' N, 130°21.54' W;
anchor: 54°19.59' N, 130°21.62' W (NAD 27)

Ryan Point Reef Light Chart 3957;
position: 54°21.56' N, 130°29.95' W (NAD 83)

**Saanich Inlet: Brentwood Bay,
Butchart Cove, Todd Inlet**
Chart 3313 (inset), p.13;
Butchart Cove dinghy dock:
48°34.07' N, 123°28.21' W (NAD 83)

Sac Bay Chart 3808;
entrance: 52°32.40' N, 131°40.10' W;
anchor: 52°31.97' N, 131°40.38' W (NAD 27)

Safe Cove Chart 3552;
position: 51°05.00' N, 126°55.00' W
(NAD 27)

Safety Cove Chart 3934;
entrance: 51°31.70' N, 127°54.50' W
anchor: 51°31.85' N, 127°55.88' W (NAD 83)

Sainty Point Light Chart 3772;
position: (on pt, opposite Yolk Pt, S.
entrance to Grenville Channel):
53°22.30' N, 129°18.67' W (NAD 27)

Salmon Arm Chart 3552 (inset);
entrance: 51°02.55′ N, 126°42.70′ W
(NAD 27)

Salmon Bay Chart 3544;
logging area entrance:
50°23.72′ N, 125°57.43′ W (NAD 83)

Salmon Bay Chart 3734;
entrance: 52°28.90′ N, 128°13.20′ W;
anchor: 52°29.12′ N, 128°12.95′ W (NAD 27)

Salmon Cove Chart 3933;
anchor (north side):
55°16.23′ N, 129°50.75′ W (NAD 27)

Salmon Inlet Chart 3312;
Kunechin Islets light:
49°37.20′ N, 123°48.16′ W (NAD 27)

Salter Point Light Chart 3664;
position: 49°40.97′ N, 126°35.14′ W
(NAD 27)

Saltery Bay Chart 3312;
public floats: 49°46.93′ N, 124°10.41′ W
(NAD 27)

Saltery Bay Chart 3663;
entrance 49°52.0′ N, 126°48.5′ W (NAD 27)

Saltery Bay Light Chart 3514;
position: 49°46.90′ N, 124°10.55′ W
(NAD 83)

Saltspring Bay Chart 3891;
anchor: 53°11.88′ N, 132°12.97′ W (NAD 83)

Saltspring Island Sector Light
Chart 3442;
position: 48°55.40′ N, 123°33.13′ W
(NAD 27)

San Carlos Point Light Chart 3664;
position: (on pt) 49°41.17′ N, 126°31.17′ W
(NAD 27)

San Jose Islets Light Chart 3671;
position: (W. extremity of westerly isl.):
48°54.10′ N, 125°03.39′ W (NAD 27)

San Josef Bay Chart 3624
entrance: 50°39.20′ N, 128°20.00′ W (NAD 27)

San Josef Inner Bay, North Side
Chart 3624;
anchor: 50°40.30′ N, 128°17.80′ W (NAD 27)

San Josef Inner Bay, South Side
Chart 3624;
anchor: 50°39.10′ N, 128°17.80′ W (NAD 27)

San Juan Point Light Chart 3647;
position: (on pt): 48°31.90′ N, 124°27.40′ W
(NAD 27)

San Mateo Bay Chart 3668;
Bernard Point float:
48°56.77′ N, 125°00.14′ W (NAD 83)

Sand Spit Light Chart Nil;
position: (end of Spit, partly on sand):
54°42.97′ N, 125°58.67′ W (NAD unknown)

Sandell Bay Chart 3932;
position: 51°39.60′ N, 127°32.87′ W (NAD 83)

Sandfly Bay Chart 3933;
anchor: 55°09.80′ N, 130°09.08′ W
(NAD 27)

"Sandford Island Cove" Chart 3671;
anchor: 48°52.42′ N, 125°09.84′ W
(NAD 27)

Sands Heads Light and Bell Buoy S1
Chart 3490;
position: (N. side of Fraser R. entrance):
49°06.19′ N, 123°18.57′ W (NAD 83)

Sands Heads Light Chart 3490;
position: (near outer end of Steveston Jetty):
49°06.37′ N, 123°18.11′ W (NAD 83)

Sandspit Aeronautical Light
Chart 3890;
position: 53°15.17' N, 131°48.80' W
(NAD 27)

Sandspit Chart 3890 (inset);
entrance: 53°14.53' N, 131°51.72' W;
position (public wharf):
53°15.26' N, 131°49.30' W (NAD 27)

Sandy Island Marine Park Chart 3527;
anchor: 49°36.98' N, 124°50.95' W (NAD 27)

Sangster Island Sector Light
Chart 3512;
position: (SW tip of isl.): 49°25.43' N,
124°11.50' W (NAD 27)

Sans Peur Passage Chart 3786;
south entrance: 51°56.14' N, 128°11.95' W;
anchor: 51°56.75' N, 128°12.46' W (NAD 27)

Sansum Narrows Charts 3313, p. 14;
position: 48°45' N, 123°34' W (NAD 83)

Santa Gertrudis Cove ("Dawleys")
Chart 3664;
south anchorage: 49°36.13' N, 126°37.19' W
(NAD 27)

Santiago Light Chart 3663;
position: 49°47.27' N, 126°39.17' W
(NAD 27)

Sapperton Bar Dyke Light Chart 3490;
position: (W. end of S. wing of dyke):
49°13.18' N, 122°50.82' W (NAD 83)

Sarah Island Light Chart 3738;
position: (N. extremity of isl.):
52°53.03' N, 128°30.55' W (NAD 27)

Sargeant Bay Chart 3311;
anchor: 49°28.50' N, 123°51.55' W (NAD 83)

Sargeaunt Passage Chart 3515;
anchor south of narrows: 50°41.55' N,
126°11.75' W;
anchor north of narrows:
50°42.02' N, 126°11.81' W (NAD 83)

Satuma Island Sector Light
Chart 3441;
position: (on East Point):
48°47.00' N, 123°02.70' W (NAD 27)

Saunders Creek Light Chart 3772;
position: (on shore opposite Saunders
Creek): 53°36.27' N, 129°41.60' W (NAD 27)

Saunders Island Light Buoy E20
Chart 3787;
position: (off W. end of isl.):
52°10.33' N, 128°07.30' W (NAD 27)

Savary Island Wharf Light Chart 3311;
position: (on wharf, N. side of isl.):
49°56.76' N, 124°46.69' W (NAD 83)

Savary Island, Keefer Bay Charts 3311,
sheet 5, 3538;
Keefer Public Wharf:
49°56.76' N, 124°46.69' W (NAD 83)

Scarlett Point Light Chart 3549;
position: (on the pt, entrance to Christie
Passage): 50°51.63' N, 127°36.67' W
(NAD 83)

Schooner Channel Chart 3921;
south entrance: 51°02.25' N, 127°31.95' W;
north entrance: 51°05.08' N, 127°31.24' W
(NAD 83)

Schooner Cove Breakwater Light
Chart 3459;
position: (extremity of breakwater):
49°17.28' N, 124°07.90' W (NAD 27)

Schooner Cove Chart 3459;
breakwater light: 49°17.28' N, 124°07.90' W
(NAD 27)

Schooner Cove Chart 3640;
anchor: 49°03.93' N, 125°48.63' W (NAD 27)

Schooner Passage Charts 3747, 3927;
south entrance: 53°44.50' N, 130°25.30' W;
north entrance: 53°47.20' N, 130°23.20' W
(NAD 27)

Schooner Reefs Light Chart 3459;
position: (W. side of Georgia Strait):
49°17.62' N, 124°07.70' W (NAD 27)

Schwartzenberg Lagoon Chart 3552;
entrance: 51°05.00' N, 127°11.75' W
(NAD 27)

Scott Cove and "West Scott Cove"
Chart 3515;
anchor Scott Cove:
50°46.05' N, 126°27.94' W;
anchor "West Scott Cove":
50°45.88' N, 126°28.20' W (NAD 83)

Scottie Bay Chart 3512;
anchor: 49°30.80' N, 124°20.57' W (NAD 27)

Scouler Entrance Chart 3651
south entrance: 50°18.33' N, 127°50.00' W
(NAD 83)

"Scow Bay" Chart 3683;
anchor: 50°06.27' N, 127°30.83' W (NAD 27)

Scroggs Rocks Light Chart 3419;
position: (S. side of rocks):
48°25.58' N, 123°26.25' W (NAD 83)

Scudder Point Light Chart 3809;
position: (on pt): 52°26.80' N, 131°14.17' W
(NAD 27)

"Sea Cave Cove" Chart 3663;
anchor: 59°51.64' N, 127°04.62' W (NAD 27)

Sea Otter Cove Chart 3624 (insert);
anchor: 50°40.74' N, 128°20.94' W (NAD 27)

Sea Otter Cove Light Chart 3624;
position: (E. side of cove):
50°40.27' N, 128°21.03' W (NAD 27)

Sea Otter Inlet Chart 3784;
entrance: 51°50.20' N, 128°00.90' W;
anchor (west): 51°50.38' N, 128°03.47' W
(NAD 27)

**Sea Otter West ODAS Light Buoy
46024** Chart 3744;
position: 51°22.50' N, 128°44.70' W
(NAD 27)

Seabird Rocks Light Chart 3602;
position: (on largest rock):
48°45.00' N, 125°09.15' W (NAD 27)

Seaforth Channel Charts 3720, 3787;
Dryad Point light:
52°11.18' N, 128°06.60' W; turning point:
52°11.20 ' N, 128°06.00' W (NAD 27)

Seal Cove Chart 3958;
entrance: 54°19.96' N, 130° 16.70' W
(NAD 27)

Seal Cove Light Chart 3958;
position: 54°20.21' N, 130°16.59' W
(NAD 83)

Seal Rocks Light Chart 3761;
position: (on highest rock of Seal Rocks
group): 54°00.00' N, 130°47.43' W (NAD 27)

**Seapool Rocks Light and Whistle
Buoy Y49** Chart 3671;
position: (on E. side of rocks):
48°48.90' N, 125°12.23' W (NAD 27)

Sechelt Chart 3311;
position: 49°28.13' N, 123°44.79' W
(NAD 83)

Sechelt Inlet Charts 3512, 3514;
entrance: 49°46.25' N, 123°57.60' W
(NAD 27)

Sechelt Islets Light Chart 3514;
position: (S. end of centre isl.):
49°44.39' N, 123°53.82' W (NAD 83)

Sechelt Rapids ("Skookumchuck Rapids") Chart 3312;
green flashing light: 49°44.39' N,
123°53.82' W (NAD 27)

Second Narrows East Light
Chart 3494;
position: (NW of Berry Point):
49°17.94' N, 122°59.80' W (NAD 83)

Second Narrows Highway Bridge (Lower) Light Chart 3494;
position: 49°17.70' N, 123°01.50' W
(NAD 83)

Second Narrows Highway Bridge (Upper) Light Chart 3494;
position: (W. side of bridge):
49°17.70' N, 123°01.50' W (NAD 83)

Second Narrows Highway Bridge Light Chart 3494;
position: (level with roadway):
49°17.68' N, 123°01.58' W (NAD 83)

Second Narrows Light Chart 3494;
position: (6.5 cables E. of railway bridge):
49°17.81' N, 123°00.48' W (NAD 83)

Secret Cove Chart 3535
(Secret Cove and Smuggler Cove inset);
anchor: 49°32.07' N, 123°58.00' W (NAD 27)

Secret Cove Chart 3909 (inset);
west entrance (Henry Island):
54°00.64' N, 130°41.37' W (NAD 27);
anchor: 54°.00.10' N, 130° 40.10' W
(NAD 83)

Secret Cove Entrance Light
Chart 3535;
position: 49°31.69' N, 123°57.93' W
(NAD 27)

"Secretary Islands Window"
Chart 3313, p.17;
anchor: 48°57.68' N, 123°35.23' W (NAD 83)

Section Cove Chart 3809;
entrance: 52°25.50' N, 131°21.70' W;
south buoy: 52°25.10' N, 131°22.70' W
(NAD 27)

Security Bay Chart 3931;
entrance: 51°22.02' N, 127°28.33' W (NAD 83)

Selby Cove Chart 3313, p. 10;
anchor: 48°50.05' N, 123°23.93' W (NAD 83)

Selma Park Breakwater Light
Chart 3311;
position: (outer end of breakwater):
49°28.00' N, 123°44.60' W (NAD 83)

Selma Park Chart 3311;
breakwater light: 49°28.00' N, 123°44.60' W
(NAD 83)

Selwyn Inlet Chart 3807;
entrance: 52°50.70' N, 131°40.00' W (NAD 27)

Selwyn Point Light Chart 3807;
position: (eastern extremity of pt):
52°51.72' N, 131°50.68' W (NAD 27)

Senanus Island Light Chart 3441;
position: (NW extremity of isl.):
48°35.58' N, 123°29.13' W (NAD 27)

Sentry Shoals ODAS Light Buoy 46131 Chart 3513;
position: 49°54.40' N, 124°59.10' W
(NAD 27)

Separation Head Light Chart 3539;
position: 50°10.74' N, 125°21.28' W
(NAD 83)

Separation Point Light Chart 3470;
position: (outer extremity of pt):
48°44.58' N, 123°34.13' W (NAD unknown)

Serpent Point Light Chart 3785;
position: 52°04.62' N, 127°59.55' W
(NAD 27)

Serpentine Inlet Charts 3761, 3927;
entrance: 53°56.02' N, 130°40.24' W;
anchor: 53°56.65' N, 130° 40.80' W (NAD 27)

Sewell Inlet Chart 3894;
entrance: 52°54.10' N, 131°53.90' W
(NAD 27)

Seymour Inlet Chart 3552
See Nakwakto Rapids.

Seymour Island Light Chart 3544;
position: (S. side of isl.):
50°28.67' N, 125°52.13' W (NAD 83)

Seymour Narrows Charts 3539, 3540
south entrance: 50°07.67' N, 125°21.17' W
north entrance: 50°09.53' N, 125°21.50' W
(NAD 83)

Seymour River Chart 3552 (inset);
position: 51°11.35' N, 126°40.00' W
(NAD 27)

Shack Bay Chart 3940;
anchor: 52°23.34' N, 127°51.62' W (NAD 83)

Shag Rock Light Chart 3892;
position: (NW of Cape Naden):
54°09.45' N, 132°39.03' W (NAD 27)

Shaman Cove Charts 3747, 3927;
entrance: 53°45.15' N, 130°25.65' W;
anchor: 53°45.54' N, 130°36.03' W (NAD 27)

Shark Cove Chart 3313 (inset), p. 12;
position: 48°45.95' N, 123°15.55' W
(NAD 83)

Sharp Point Light Chart 3643;
position: (E. side entrance, Hot Springs
Cove): 49°20.87' N, 126°15.50' W
(NAD unknown)

"Shaw Point Cove" Chart 3544;
position: 50°28.63' N, 125°54.67' W (NAD 83)

Shawl Bay Chart 3515;
anchor: 50°50.80' N, 126°33.80' W (NAD 83)

Shears Islands Light Chart 3670;
position: 48°59.99' N, 125°19.35' W
(NAD 83)

Shearwater Charts 3785 (inset), 3787;
float: 52°08.82' N, 128°05.20' W (NAD 27)

Shearwater Passage (Mystery Reef)
Chart 3311, sheet 5;
Grant Reef light buoy "QM" at south
extremity of reefs:
49°52.08' N, 124°45.97' W;
Mystery Reef buoy "Q25", close northeast of
reef: 49°54.77' N, 124°42.72' W (NAD 83)

Sheep Passage Charts 3962, 3738;
west entrance: 52°48.60' N, 128°24.60' W;
east entrance: 52°51.00' N, 128°09.20' W
(NAD 27)

Shelter Bay North Chart 3548;
entrance: 50°58.21' N, 127°26.73' W;
anchor: 50°58.08' N, 127°25.33' W (NAD 83)

Shelter Inlet Chart 3648
west entrance: 49°24.00' N, 126°13.60' W
(NAD 27)

"Shelter Shed" Charts 3680, 3683
Shed 1–4 southeast side Brooks Peninsula

Sheringham Point Light Chart 3606;
position: (on pt): 48°22.62' N, 123°55.19' W
(NAD 27)

"Shibasha Cove" Charts 3747, 3927;
entrance: 53°46.80' N, 130°28.83' W;
anchor: 53°46.98' N, 130°24.43' W (NAD 27)

Shingle Bay Chart 3313, p. 8;
anchor: 48°46.87' N, 123°18.65' W (NAD 83)

Shingle Spit Chart 3527, 3513;
anchor: 49°30.70' N, 124°42.35' W (NAD 27)

Ship Anchorage Chart 3772;
entrance: 53°41.20' N, 129°43.90' W
(NAD 27)

Shkgeaum Bay Sector Light
Chart 3955;
position: (E. side of bay):
54°18.51' N, 130°24.37' W (NAD 27)

Shoal Bay Chart 3543;
anchor: 50°27.55' N, 125°21.80' W (NAD 27)

Shoal Harbour Chart 3515;
anchor (east): 50°44.08' N, 126°29.50' W
(NAD 83)

Shoal Point Light Chart 3415;
position: (W. of pt):
48°25.42' N, 123°23.26' W (NAD 27)

Shoal Point Light Chart 3490;
position: (E. end of trifurcation wall):
49°11.75' N, 122°54.93' W (NAD 83)

"Shorter Point Cove" Chart, 3543;
position: 50°24.34' N, 125°42.76' W (NAD 83)

Shotbolt Bay Chart 3932;
position: 51°39.23' N, 127°21.28' W (NAD 83)

Shrub Island Light Chart 3955;
position: (W. end of Venn Passage):
54°19.83' N, 130°27.56' W (NAD 27)

Shushartie Bay Chart 3549;
position: 50°51.27' N, 127°51.36' W (NAD 83)

Shute Reef Light Chart 3476;
position: (on rock):
48°42.63' N, 123°25.84' W (NAD 27)

Shuttleworth Bight Chart 3624;
entrance: 50°51.65' N, 128°08.20' W
(NAD 27)

Sidney Bay Charts 3555 (inset), 3543;
position: 50°31.00' N, 125°36.40' W (NAD 27)

Sidney Breakwater Light Chart 3476;
position: (S. extremity of breakwater):
48°39.15' N, 123°23.52' W (NAD 27)

Sidney Breakwater Light Chart 3476;
position: (on wharf):
48°38.60' N, 123°23.73' W (NAD 27)

Sidney Channel Light Buoy U2
Chart 3441;
position: (SW of western shoal, N. entrance
to channel): 48°37.47' N, 123°20.70' W
(NAD 27)

Sidney Charts 3313, p. 7;
breakwater light: 48°39.22' N, 123°23.43' W
Chart 3476 (NAD 83)

Sidney Spit Light Chart 3476;
position: (extreme NW end of spit):
48°39.24' N, 123°20.68' W (NAD 27)

Sidney Spit Marine Park
Chart 3313, p. 6;
float: 48°38.50' N, 123°19.93' W (NAD 83)

Silva Bay Chart 3313, p. 19;
anchor 49°09.05' N, 123°41.78' W (NAD 83)

Silva Bay Light Chart 3475;
position: (NW of Tugboat Isl.):
49°09.22' N, 123°41.45' W (NAD 27)

Silvester Bay Chart 3550;
position (north end):
51°09.45' N, 127°45.30' W (NAD 83)

Simoom Sound Chart 3515 (inset);
entrance: 50°47.40' N, 126°30.75' W (NAD 83)

**Sir Alexander Mackenzie Rock
Provincial Park** Chart 3781;
anchor: 52°22.69' N, 127°28.22' W (NAD 27)

Sir Edmund Bay Chart 3515;
entrance: 50°50.15' N, 126°35.30' W (NAD 83)

Sisters Islets Light Chart 3513;
position: (on easterly and largest rock):
49°29.22' N, 124°26.00' W (NAD 27)

Skaat Harbour Chart 3809; north
entrance: 52°26.40' N, 131°24.70' W;
anchor: 52°23.05' N, 131°26.33' W (NAD 27)

Skedans Bay Chart 3894; north
entrance: 52°57.50' N, 131°35.40' W;
position: 53°56.75' N, 131°37.55' W (NAD 27)

Skene Cove Charts 3773, 3927;
entrance: 53°50.90' N, 130°20.25' W (NAD 27)

Skerry Bay Chart 3512;
anchor 49°29.98' N, 124°13.94' W (NAD 27)

Skiakl Bay Chart 3956;
entrance: 54°07.30' N, 130°46.90' W;
position (northeast basin):
54°08.80' N, 130°46.50' W (NAD 83)

Skidegate Channel Chart 3891; east
entrance: 53°09.90' N, 132°08.60' W;
west entrance: 53°09.80' N, 132°34.50' W
(NAD 83)

**Skidegate Inlet to Lawn and Rose
Points** Charts 3902, 3802;
position (Lawn Point Buoy "C14"):
53°25.54' N, 131°53.10' W (NAD 27)

Skidegate Landing Chart 3890;
entrance: 53°14.70' N, 132°00.50' W (NAD 27)

Skookum Island Light Chart 3514;
position: (NW end of isl.):
49°43.53' N, 123°52.65' W (NAD 83)

Skookumchuck Narrows Light
Chart 3514;
position: (on drying rock):
49°45.60' N, 123°55.97' W (NAD 83)

Skowquiltz Bay Chart 3730;
position: 52°35.80' N, 127°09.20' W
(NAD 27)

Skull Cove Chart 3921;
entrance: 51°02.95' N, 127°33.30' W;
anchor (north): 51°03.23' N, 127°33.68' W
(NAD 83)

Sleepy Bay Chart 3934;
entrance: 51°30.48' N, 127°40.13' W;
position: 51°30.14' N, 127°39.23' W
(NAD 83)

Slim Inlet Chart 3809;
entrance: 52°17.85' N, 131°19.35' W;
anchor: 52°17.08' N. 131°19.23' W
(NAD 27)

Slingsby Channel Chart 3921;
east entrance (0.15 mile South Kitching
Point): 51°05.50' N, 127°31.30' W (NAD 83);
Chart 3551;
west entrance: 51°05.00' N, 127°39.05' W
(NAD 27)

Slippery Rock Light Chart 3959;
position: (on rock):
54°23.97' N, 130°29.72' W (NAD 83)

Sloop Islet Light Chart 3893;
position: (on islet):
53°45.50' N, 132°14.57' W (NAD 27)

Small Inlet Chart 3539;
anchor: 50°14.48' N, 125°17.13' W (NAD 83)

Smelt Bay Charts 3538, 3312;
position: 50°02.00' N, 124°59.80' W
(NAD 83)

Smith Cove Chart 3679;
anchor: 50°29.09' N, 127°35.13' W (NAD 83)

Smith Inlet Chart 3931;
entrance: 51°18.90' N, 127°32.30' W
(NAD 83)

"Smithers Island Cove"
Chart 3719 (inset);
entrance: 52°45.37' N, 129°04.68' W;
anchor: 52°45.53' N, 129°04.14' W (NAD 27)

Smuggler Cove and Marine Park
Chart 3535
(Secret Cove and Smuggler Cove inset);
anchor: 49°30.78' N, 123°57.64' W (NAD 27)

Snake Island Light Chart 3458;
position: (N. end of isl.):
49°13.05' N, 123°53.35' W (NAD 83)

Snake Island Reef Light and Bell Buoy P2 Chart 3458;
position: (E. of isl.):
49°12.50' N, 123°53.05' W (NAD 83)

Snug Basin Chart 3646 (inset);
anchor: 49°01.20' N, 125°01.72' W (NAD 83)

Snug Cove Chart 3534 (inset);
anchor: 49°22.74' N, 123°19.79' W (NAD 83)

Snug Cove Light Chart 3534;
position: (on pt at entrance to cove):
49°22.81' N, 123°19.25' W (NAD 83)

Snug Cove North Light Chart 3534;
position: 49°22.88' N, 123°19.49' W
(NAD 83)

Snuggery Cove and Port Renfrew
Chart 3647;
anchor: 48°33.32' N, 124°25.11' W (NAD 27)

"Soda Waterfalls" Chart 3312;
position: 49°59.95' N, 123°56.60' W (NAD 27)

Sointula Breakwater Light Chart 3546;
position: (outer end of breakwater):
50°38.46' N, 127°01.93' W (NAD 83)

Sointula Chart 3546;
public float: 50°37.62' N, 127°01.13' W
(NAD 83)

Solander Island Chart 3680;
north way-point: 50°07.07' N, 127°57.27' W;
south way-point: 50°06.37' N, 127°57.16' W
(NAD 27)

Solander Island Light Chart 3680;
position 50°06.68' N, 127°56.28' W (NAD 27)

Somass River Light Chart 3668;
position: 49°14.66' N, 124°49.46' W (NAD 83)

Somerville Bay Chart 3994;
entrance: 54°47.75' N, 130°12.40' W;
anchor: 54°46.50' N, 130°13.48' W (NAD 27)

Sooke Basin Charts 3430, 3641;
anchor: 48°22.35' N, 123°40.77' W
(NAD unknown)

Sooke Harbour Inner Range Light
Chart 3430;
position: 48°21.77' N, 123°42.50' W
(NAD unknown)

Sooke Harbour Outer Range Light
Chart 3430;
position: 48°21.62' N, 123°42.32' W
(NAD unknown)

Sooke Harbour Range Light
Chart 3430;
position: 48°21.92' N, 123°43.67' W
(NAD unknown)

Sooke Inlet Charts 3430, 3641;
Whiffin Spit light: 48°21.52' N, 123° 42.61' W
Whiffin Spit anchor: 48°21.62' N, 123°43.05' W
(NAD unknown)

"South Arm," Sea Otter Inlet
Chart 3784;
anchor (South Arm):
51°49.54' N, 128°01.59' W (NAD 27)

South Bay Light Chart 3890;
position: (W. end of small isl.):
53°09.88' N, 132°03.90' W (NAD 27)

South Bentinck Arm Chart 3730;
entrance: 52°18.50' N, 126°59.00' W
(NAD 27)

South Cod Reef South Cardinal Light Buoy US Chart 3441;
position: 48°39.00' N, 123°18.00' W
(NAD 27)

"South Congreve Cove" Chart 3671;
entrance: 48°55.34' N, 125°01.40' W
(NAD 27)

"South Cove" Chart 3683;
anchor west nook: 50°05.94' N, 127°31.47' W
(NAD 27)

South Moresby ODAS Light Buoy 46147 Chart 3002;
position: 51°49.30' N, 131°12.10' W
(NAD 27)

South Nomad ODAS Light Buoy 46036 Chart 3000;
position: 48°21.20' N, 133°55.30' W (NAD 27)

South Rachael Island Light
Chart 3957;
position: (S. end of isl.):
54°11.83' N, 130°33.44' W (NAD 83)

"Southeast Lagoon Cove" Chart 3940;
anchor: 52°28.24' N, 127°56.22' W
(NAD 83)

Southey Bay, Saltspring Island
Chart 3313, pp. 15, 17;
anchor: 48°56.63' N, 123°35.72' W (NAD 83)

Southey Point Light Chart 3442;
position: (N. extremity of Saltspring Isl.):
48°56.75' N, 123°35.72' W (NAD 27)

Southgate Island
Chart 3921 (inset) or 3551;
south entrance: 51°00.15' N, 127°32.25' W;
anchor: 51°00.94' N, 127°31.37' W (NAD 83)

Spanish Bank Anchorage East Light
Chart 3481;
position: 49°17.12' N, 123°13.07' W (NAD 27)

Spanish Bank Light Chart 3481;
position: 49°17.25' N, 123°13.52' W
(NAD 27)

Spanish Banks No. 1 Light Chart 3481;
position: (entrance to Burrard Inlet):
49°16.70' N, 123°15.63' W (NAD 27)

Spanish Banks No. 2 Light Chart 3481;
position: (entrance to English Bay):
49°17.37' N, 123°14.80' W (NAD 27)

Spencer Cove Chart 3679;
entrance: 50°30.03' N, 127°52.53' W
(NAD 83)

Spencer Creek Light Chart 3668;
position: 48°59.33' N, 124°53.50' W
(NAD 83)

"Spicer Island Complex" ("Spicer Anchorage") Charts 3747, 3927;
entrance (southwest):
53°44.7' N, 130°21.71' W;
anchor (unnamed island nook):
53°44.88' N, 130°21.57' W (NAD 27)

Spider Anchorage Chart 3784;
Edna Island anchor:
51°49.52' N, 128°14.38' W (NAD 27)

Spider Channel Light Chart 3786;
position: (N. end of Spider Isl.):
51°51.77' N, 128°15.45' W (NAD 27)

Spiller Channel Chart 3940;
south entrance: 52°17.00' N, 128°14.20' W
(NAD 83)

Spiller Inlet Chart 3940;
entrance: 52°30.70' N, 128°04.60' W;
anchor: 52°38.92' N, 128°03.03' W
(NAD 83)

Spilsbury Point Light Chart 3311;
position: (N. tip of Hernado Isl.):
50°00.21' N, 124°56.57' W (NAD 83)

Spire Ledge Light and Bell Buoy D47 Chart 3958;
position: (off E. end of ledge):
54°14.87' N, 130°21.13' W (NAD 83)

Spitfire Channel Charts 3784, 3786;
east entrance: 51°50.69' N, 128°10.07' W;
west entrance: 51°51.64' N, 128°13.78' W (NAD 27)

"Spitfire East Cove" Charts 3784, 3786;
anchor: 51°51.59' N, 128°11.47' W (NAD 27)

"Spitfire Lagoon" Chart 3786;
entrance: 51°51.92' N, 128°12.59' W;
anchor: 51°52.10' N, 128°12.66' W (NAD 27)

"Spitfire West Cove" Chart 3784;
anchor: 51°51.44' N, 128°12.21' W (NAD 27)

Split Head (Separation Point) Chart 3734;
anchor: 52°40.38' N, 128°33.18' W (NAD 27)

Split Head (Separation Point) Light Chart 3711;
position: 52°40.65' N, 128°32.85' W (NAD 27)

"Spout Islet Cove" Chart 3546;
anchor: 50°35.15' N, 126°45.00' W (NAD 83)

"Spouter Island Anchorage" Chart 3664;
anchor: 49°37.40' N, 126°32.97' W (NAD 27)

Spring Bay, Spanish Cove Chart 3513;
anchor: 49°31.30' N, 124°21.73' W (NAD 27)

Spring Cove Chart 3646;
anchor: 48°55.64' N, 125°31.74' W (NAD 83)

Spring Cove Light Buoy Y45 Chart 3646;
position: (SE of Hyphocus Isl.):
48°55.65' N, 125°31.43' W (NAD 83)

Spring Passage Chart 3546;
east entrance: 50°38.50' N, 126°34.00' W (NAD 83)

Sproat Bay Chart 3671;
anchor: 48°54.36' N, 125°04.56' W (NAD 27)

Squall Bay Charts 3753 (inset), 3746;
south entrance: 53°32.95' N, 130°06.20' W;
anchor (north): 53°34.02' N, 130°07.26' W (NAD 27)

Squally Channel Chart 3742;
south entrance: 53°04.10' N, 129°18.10' W (NAD 27)

Squamish Approach Light Chart 3534;
position: (entrance to channel):
49°40.80' N, 123°09.83' W (NAD 83)

Squamish Chart 3534;
channel entrance light:
49°40.80' N, 123°09.83' W (NAD 83)

Squamish Range Light Chart 3534;
position: (E. bank of Squamish River):
49°41.52' N, 123°09.12' W (NAD 83)

Squamish Terminal No. 1 Light Chart 3534;
position: 49°41.01' N, 123°10.49' W (NAD 83)

Squamish Terminal No. 2 Light Chart 3534;
position: 49°40.94' N, 123°10.29' W (NAD 83)

Squamish Terminal Range Light Chart 3534;
position: 49°41.21' N, 123°10.30' W (NAD 83)

Squirrel Cove, Cortes Island
Chart 3555;
anchor: 50°08.52' N, 124°55.05' W (NAD 27)

Squitty Bay Chart 3512;
entrance: 49°26.60' N, 124°09.60' W
(NAD 27)

St. John Harbour
Chart 3711 (inset), 3787;
entrance: 52°12.30' N, 128°28.50' W
(NAD 27)

St. John Lagoon Chart 3711 (inset);
entrance: 52°11.02' N, 128°27.90' W;
anchor (first basin):
52°10.64' N, 128°27.28' W (NAD 27)

St. Vincent Bay, Sykes Island
Chart 3312;
anchor: 49°48.68' N, 124°04.85' W (NAD 27)

St. Vincent Bight Chart 3545;
position: 50°27.47' N, 126°10.15' W (NAD 83)

Stalkungi Cove Chart 3807;
entrance: 52°45.70' N, 131°44.55' W;
anchor: 52°46.00' N, 131°44.80' W (NAD 27)

Stamp Narrows Light Chart 3668;
position: (W. side of narrows):
49°11.00' N, 124°49.30' W (NAD 83)

Staniforth Point Light Chart 3743;
position: (on pt): 53°34.15' N, 128°48.90' W
(NAD 27)

Star Point Light Chart 3668;
position: (on pt, S. shore):
48°58.62' N, 124°56.83' W (NAD 83)

Starling Point Light Chart 3648;
position: (on pt): 49°23.64' N, 126°13.75' W
(NAD 27)

Steamboat Bay Chart 3547 (inset);
anchor: 50°56.18' N, 126°48.24' W (NAD 83)

Steamer Cove Chart 3648;
anchor: 49°22.55' N, 126°11.45' W
(NAD 27)

Steamer Passage Chart 3994
south entrance: 54°41.30' N, 130°22.35' W;
north entrance: 54°47.50' N, 130°11.30' W
(NAD 27)

Steamer Point Light Chart 3663;
position: (on pt): 49°53.20' N, 126°47.80' W
(NAD 27)

Steep Head Chart 3545;
anchor: 50°40.30' N, 126°11.00' W (NAD 83)

Steep Island Light Chart 3540;
position: (W. side of isl.):
50°04.75' N, 125°15.20' W (NAD 83)

**Stenhouse Shoal Light and Whistle
Buoy D59** Chart 3957;
position: 54°20.12' N, 130°56.04' W
(NAD 83)

Stephens Narrows, Leavitt Lagoon
Chart 3721;
entrance: 53°20.10' N, 129°40.82' W
(NAD 27)

Stephens Passage Chart 3956;
position (drying passage):
54°07.10' N, 130°08.80' W (NAD 83)

Steveston Breakwater Light
Chart 3490;
position: (W. extremity of breakwater):
49°07.47' N, 123°11.73' W (NAD 83)

Steveston Jetty Chart 3490;
Sand Heads light: 49°06.19' N,
123°18.57' W (NAD 83)

Stewart Chart 3794;
Stewart Yacht Club public float:
55°55.03' N, 130°00.52' W (NAD 27)

Stewart Dolphin West Light
Chart 3794;
position: 55°54.79' N, 130°00.18' W
(NAD 27)

Stewart Inlet Chart 3786;
entrance: 51°52.50' N, 128°07.68' W (NAD 27)

Stewart Light Chart 3794;
position: 55°54.65' N, 129°59.37' W
(NAD 27)

Stimpson Reef Light Chart 3545;
position: 50°29.73' N, 126°12.14' W
(NAD 83)

Stockham Island Light Chart 3685;
position: (off W. end of isl.):
49°10.30' N, 125°54.37' W (NAD 83)

Storm Bay Chart 3312;
anchor: 49°39.80' N, 123°49.58' W (NAD 27)

"Stormy Bay" Chart 3743;
entrance: 53°46.50' N, 128°48.20' W;
anchor: 53°46.39' N, 128°48.17' W (NAD 27)

Story Point Light Chart 3787;
position: (on pt): 52°08.92' N, 128°08.02' W
(NAD 27)

Strachan Bay Chart 3552;
entrance: 51°10.00' N, 127°25.30' W;
anchor (cove): 51°09.58' N, 127°28.43' W
(NAD 27)

Straggling Islands Light Chart 3679;
position: (W end of largest isl.):
50°35.67' N, 127°42.14' W (NAD 83)

Striae Islands Light Chart 3895;
position: (easterly islet of Striae Isl. group):
54°05.22' N, 132°14.80' W (NAD 27)

Strombeck Bay Chart 3920;
entrance: 55°22.90' N, 129°46.90' W;
anchor: 55°22.42' N, 129°47.00' W (NAD 83)

"Stryker Island Nook" Chart 3787;
entrance: 52°04.50' N, 128°20.40' W;
anchor: 52°06.02' N, 128°20.54' W (NAD 27)

Stuart Anchorage Chart 3773;
entrance: 53°51.56' N, 130°04.27' W;
anchor (westernmost cove):
53°51.10' N, 130°04.46' W (NAD 27)

Stuart Bay Chart 3646;
anchor: 48°56.06' N, 125°31.33' W (NAD 83)

Stuart Narrows Chart 3547;
west entrance: 50°53.65' N, 126°54.60' W
(NAD 83)

Stubbs Spit Light Buoy Y25
Chart 3685;
position: (NE extremity of sand bank N. of
isl.): 49°10.12' N, 125°54.43' W (NAD 83)

Stumaun Bay Chart 3963;
position: 54°33.70' N, 130°23.44' W
(NAD 83)

Sturdies Bay, Galiano Island
Chart 3313, p. 9;
public float: 48°52.62' N, 123°18.93' W
(NAD 83)

Sturgeon Bank Light Buoy T10
Chart 3463;
position: 49°11.00' N, 123°17.70' W
(NAD 27)

Subtle Islands Charts 3538, 3312;
position east side of sand spit:
50°07.14' N, 125°04.61' W (NAD 27)

Sue Channel Chart 3743;
east entrance: 53°43.00' N, 128°50.00' W;
west entrance: 53°40.80' N, 128°04.80' W
(NAD 27)

Sullivan Bay Chart 3547;
float: 50°53.15' N, 126°49.58' W (NAD 83)

Sulphur Arm Chart 3921;
entrance: 51°40.50' N, 127°45.30' W
(NAD 83)

Sulphur Passage Chart 3648;
position: 49°24' N, 126°04' W (NAD 27)

Summers Bay Chart 3552;
position: 51°10.20' N, 127°01.50' W (NAD 27)

Sunday Harbour Chart 3546, 3515;
anchor: 50°43.45' N, 126°41.90' W (NAD 83)

Sunny Island Light Chart 3720;
position: (rock S. of isl., Fisher Channel):
52°11.28' N, 127°52.06' W (NAD 27)

Sunshine Bay Chart 3934;
entrance: 51°28.55' N, 127°40.75' W;
anchor: 51°28.65' N, 127°40.00' W (NAD 83)

Superstition Point Chart 3786;
turning point (0.2 mile WNW of
Superstition Point):
51°53.47' N, 128°15.56' W (NAD 27)

"Superstition Point Cove" Chart 3786;
entrance: 51°53.67' N, 128°14.79' W
(NAD 27)

Surf Inlet Chart 3737;
entrance: 52°53.00' N, 129°08.50' W
(NAD 27)

"Surf Inlet Head" Chart 3737;
position: 53°01.75' N, 128°59.20' W
(NAD 27)

Surge Narrows Settlement
Charts 3537 (inset), 3539;
public float: 50°13.62' N, 125°06.63' W
(NAD 27)

Surge Narrows, Beazley Passage
Charts 3537 (inset), 3539;
Beazley Narrows position:
50°13.57' N, 125°08.53' W (NAD 27)

Surgeon Islets Light Chart 3547;
position: (N. side of northernmost isl. of
group): 50°53.53' N, 126°52.23' W (NAD 83)

Surrey Islands Chart 3535
(Welcome Passage inset); position:
49°30.28' N, 123°58.80' W (NAD 27)

"Susan Islets Cove"
Charts 3559, 3312, 3538;
anchor: 50°04.25' N, 124°42.13' W (NAD 27)

Susan Rock Light Chart 3728;
position: (on the rock, Milbanke Sound):
52°17.20' N, 128°30.25' W (NAD 27)

Sutherland Bay Chart 3547;
anchor: 50°55.67' N, 127°10.95' W (NAD 83)

Sutil Point Light and Bell Buoy Q20
Chart 3594;
position: (SW of pt):
50°00.15' N, 125°00.48' W (NAD unknown)

Sutlej (Deadman) Point Light
Chart 3730;
position: (on pt): 52°22.60' N, 126°48.20' W
(NAD 27)

Sutton Islets Chart 3312;
position: 49°45.78' N, 123°56.30' W
(NAD 27)

**Sutton Rock Bifurcation Light
Buoy YH** Chart 3646;
position: (SE of the rock):
48°56.43' N, 125°32.18' W (NAD 83)

Swale Rock Light Chart 3670;
position: 48°55.56' N, 125°13.23' W
(NAD 83)

Swanson Bay Charts 3738, 3739;
entrance: 53°00.60' N, 128°30.90' W;
anchor (south): 53°00.82' N, 128°30.28' W;
anchor (north): 53°01.93' N, 128°30.77' W
(NAD 27)

Swartz Bay Chart 3313, p. 7;
public wharf: 48°41.25' N, 123°24.43' W
(NAD 83)

Swiftsure Bank Chart 3602;
position: 48°33' N, 125°00' W (NAD 27)

Swordfish Bay Chart 3786;
entrance: 51°52.56' N, 128°14.23' W (NAD 27)

Sydney Inlet Chart 3648
entrance: 49°20.90' N, 126°14.70' W (NAD 27)

**Sydney Inlet Light and Whistle Buoy
MH** Chart 3648;
position: (entrance to inlet, SE of Sharp Pt):
49°20.00' N, 126°15.52' W (NAD 27)

Sylvester Bay Chart 3920;
entrance: 55°23.20' N, 129°47.20' W;
anchor: 55°21.74' N, 129°48.42' W (NAD 83)

Tahsis Chart 3665;
public dock: 49°54.70' N, 126°39.62' W
(NAD 27)

Tahsis Inlet Charts 3664, 3663;
south entrance: 49°41.15' N, 126°35.00' W
(NAD 27)

Tahsis Narrows Chart 3663;
entrance: 49°52.00' N, 126°42.60' W
(NAD 27)

Tahsis Narrows Light Chart 3663;
position: (on Mozino Pt, E. entrance to narrows): 49°51.57' N, 126°40.35' W (NAD 27)

Tahsis Narrows North Light
Chart 3663;
position: (on NE side of narrows):
49°51.89' N, 126°42.39' W (NAD 27)

Tahsis Narrows South Light
Chart 3663;
position: (S. side of narrows): 49°51.61' N,
126°41.42' W (NAD 27)

Tahsish Inlet Chart 3682
entrance: 50°05.15' N, 127°10.25' W
(NAD 27)

Takush Harbour Charts 3934, 3931;
entrance: 51°17.26' N, 127°37.00' W
(NAD 83)

Tallac Bay Chart 3543;
anchor: 50°26.68' N, 125°28.49' W (NAD 27)

Tallheo Hot Springs Chart 3730;
position: 52°12.10' N, 126°56.05' W
(NAD 27)

Tankeeah River Chart 3940;
anchor: 52°17.74' N, 128°15.70' W
(NAD 83)

Target Bay Chart 3784;
position: 51°49.47' N, 128°01.87' W
(NAD 27)

Tasu Narrows Light Chart 3859;
position: 52°44.65' N, 132°06.40' W
(NAD 27)

Tasu Sound Light Chart 3859;
position: (on pt inside entrance, S. side):
52°44.95' N, 132°05.68' W (NAD 27)

Tate Cove Chart 3723 (inset);
entrance (Turtish Harbour):
52°44.27' N, 129°16.68' W;
anchor: 52°44.02' N, 129°16.53' W (NAD 27)

Tatnall Reefs Chart 3549
position: 50°52.50' N, 127°59.00' W (NAD 83)

Tattenham Ledge Light Buoy Q51
Chart 3535;
position: (N. extremity of ledge):
49°31.13' N, 123°59.03' W (NAD 27)

Taylor Bay Chart 3934;
anchor: 51°30.78' N, 127°36.25' W (NAD 83)

Tcenakun Point Light Chart 3891;
position: (on Rocky Pt, S. side of
entrance to Skidgate Channel):
53°09.11' N, 132°35.03' W (NAD 83)

Teakerne Arm Charts 3538, 3541;
anchor: 50°11.93' N, 124°50.80' W (NAD 27)

Telegraph Cove Chart 3546;
entrance: 50°32.88' N, 126°50.04' W
(NAD 83)

Telegraph Harbour Charts 3313, p. 16;
entrance: 48°58.00' N, 123°39.90' W;
anchor: 48°58.78' N, 123°40.23' W (NAD 83)

Telegraph Passage
Charts 3773, 3927, 3717;
south entrance: 53°55.30' N, 130°05.60' W
(NAD 27)

Telescope Passage Chart 3312;
position: 49°45.35' N, 124°08.82' W (NAD 27)

Templar Channel Light Buoy Y2
Chart 3685;
position: (NW of Tonquin Isl.):
49°08.00' N, 125°55.52' W (NAD 83)

Ten Mile Point Light Chart 3668;
position: (on pt opposite Nahmint Bay):
49°03.57' N, 124°50.37' W (NAD 83)

Tenas Island Light Chart 3711;
position: 52°42.48' N, 128°33.13' W (NAD 27)

Tenedos Bay Charts 3312, p. 9, 3538;
anchor northeast corner:
50°07.37' N, 124°41.41' W (NAD 27)

Terminal Dock East Light Chart 3493;
position: 49°17.65' N, 123°02.97' W
(NAD 83)

Texada Island Light Chart 3512;
position: (at Partington Point):
49°31.72' N, 124°13.50' W (NAD 27)

The Blow Hole Chart 3564;
narrows: 50°36.39' N, 126°18.68' W (NAD 83)

The Gorge Chart 3538, 3311;
entrance: 50°05.00' N, 125°00.95' W
(NAD 83)

"The Sanctuary" Chart 3729;
position: 52°13.50' N, 127°21.40' W (NAD 27)

"The Watchmen," Fisher Point
Chart 3787;
position: 52°15.85' N, 128°20.80' W
(NAD 27)

Theodosia Inlet Charts 3559, 3312, 3538;
entrance narrows: 50°04.13' N,
124°41.88' W (NAD 27)

Thieves Bay Chart 3313, p. 8;
position: 48°46.30' N, 123°18.80' W
(NAD 83)

Third Light Chart 3491;
position: (S. side of channel):
49°14.37' N, 123°14.80' W (NAD 83)

Thistle Passage Chart 3737;
entrance (south): 52°36.20' N, 128°45.85' W;
entrance (north): 52°39.50' N, 128°45.30' W
(NAD 27)

Thompson Bay Chart 3787;
south entrance: 52°06.20' N, 128°24.60' W
(NAD 27)

Thompson Sound Chart 3515;
anchor: 50°48.03' N, 126°00.90' W
(NAD 83);
(note: anchorage reported to be 0.4 mile
north of Sackville Island)

Thors Cove Charts 3559, 3312, 3538,
anchor: 50°03.47' N, 124°42.22' W (NAD 27)

"Thrasher Cove" Chart 3647;
anchor: 48°33.47' N, 124°28.13' W (NAD 27)

Thrasher Rock Light Chart 3475;
position: (NE extremity of Gabriola Reefs):
49°09.01' N, 123°38.42' W (NAD 27)

Thulin Passage Light Chart 3538;
position: (N. end of passage):
50°01.83' N, 124°49.55' W (NAD 83)

Thunder Bay Chart 3312;
anchor: 49°45.77' N, 124°16.05' W (NAD 27)

Thurston Bay Charts 3543, 3539;
anchor lagoon: 50°21.64' N, 125°19.07' W
(NAD 27)

Thurston Harbour
Charts 3811 (inset), 3894;
entrance: 52°50.40' N, 131°43.00' W;
anchor: 52°50.50' N, 131°44.88' W
(CHS states NAD unknown)

Tibbs Islet Light Chart 3648;
position: (on islet, S. side of Russell Channel):
49°13.78' N, 126°06.50' W (NAD 27)

Tilbury Range 1 Light Chart 3490;
position: (near N. bank of river, opposite
Tilbury Isl.): 49°08.63' N, 123°02.99' W
(NAD 83)

Tilbury Range 2 Light Chart 3490;
position: (near N. bank of river):
49°08.97' N, 123°02.26' W (NAD 83)

Tlupana Inlet Chart 3664
entrance: 49°41.65' N, 126°31.00' W (NAD 27)

Toba Inlet Chart 3541
entrance: 50°19.00' N, 124°46.30' W (NAD 27)

Tofino Breakwater Light Chart 3685;
position: (on outer end of new breakwater):
49°09.28' N, 125°54.00' W (NAD 83)

Tofino Chart 3685;
public wharf: 49°09.29' N, 125°54.53' W
(NAD 83)

Tofino East Breakwater Light
Chart 3685;
position: (on N. end of breakwater):
49°09.27' N, 125°53.95' W (NAD 83)

Tofino Inlet Chart 3649
entrance: 49°07.50' N, 125°43.80' W (NAD 27)

Tolmie Channel Charts 3734, 3738;
south entrance: 52°38.60' N, 128°31.75' W
(NAD 27)

Tolmie Channel Sector Light
Chart 3711;
position: 52°41.35' N, 128°32.64' W
(NAD 27)

Tom Bay Chart 3728;
entrance: 52°24.30' N, 128°16.10' W;
anchor: 52°23.97' N, 128°15.95' W (NAD 27)

Tom Island Light Chart 3772;
position: (on isl.): 53°32.60' N, 129°36.20' W
(NAD 27)

Tom Point Light Chart 3441;
position: (on small islet E. of pt):
48°39.75' N, 123°16.33' W (NAD 27)

"Tom's Cove" Charts 3934, 3931;
anchor: 51°18.25' N, 127°34.50' W (NAD 83)

Tombstone Bay Chart 3393; float
position: 55°24.38' N, 130°33.13' W
(NAD 27)

Tomkinson Point Light Chart 3742;
position: (on pt, Ursula Channel):
53°25.83' N, 128°54.15' W (NAD 27)

Tongass Passage Chart 3960;
south entrance: 54°43.50' N, 130°37.70' W
(NAD 83)

Topaze Harbour Chart 3544 (inset);
Jackson Bay Landing buoys:
50°31.15' N, 125°45.30' W (NAD 83)

Toquart Bay Chart 3670;
Snowden Island anchor:
49°01.60' N, 125°19.67' W (NAD 83)

Toquart Bay Light Chart 3670;
position: 49°00.67' N, 125°20.32' W
(NAD 83)

Totem Inlet Charts 3747, 3927;
entrance: 53°45.38' N, 130°24.89' W;
anchor: 53°45.75' N, 130°25.48' W (NAD 27)

Towry Point East Cove Chart 3552;
anchor: 51°03.71' N, 126°53.74' W (NAD 27)

Towry Point West Cove Chart 3552;
anchor: 51°03.71' N, 126°54.19' W (NAD 27)

Tracey Harbour, Napier Bay
Chart 3547;
anchor: 50°50.98' N, 126°51.07' W (NAD 83)

Trahey Inlet Chart 3737;
entrance (east): 52°39.90' N, 128°47.55' W
(NAD 27)

Trail Bay Chart 3963;
entrance: 54°34.28' N, 130°20.22' W;
anchor: 54°35.31' N, 130°22.25' W (NAD 83)

"Trail Creek Cove" Chart 3682;
anchor: 50°02.6' N, 127°128.9' W (NAD 27)

Trail Islands Chart 3311;
position: 49°27.53' N, 123°48.58' W
(NAD 83)

Tranquil Inlet Chart 3649;
entrance: 49°11' N, 125°41' W (NAD 27)

"Tranquilito Cove" Chart 3649;
anchor: 49°12.23' N, 125°39.93' W (NAD 27)

Treadwell Bay Chart 3921 (inset);
entrance: 51°05.77' N, 127°32.00' W;
anchor: 51°06.27' N, 127°32.68' W (NAD 83)

Tree Bluff Light and Bell Buoy D86
Chart 3963;
position: (end of shoal W of bluff):
54°25.77' N, 130°30.75' W (NAD 83)

Tree Nob Group Chart 3957;
position: 54°15.19' N, 130° 48.53' W
(NAD 83)

Trevenen Bay
Charts 3559 (inset), 3312, 3538;
anchor: 50°00.94' N, 124°44.02' W
(NAD 27)

Trevor Channel Chart 3671;
entrance: 48°50' N, 125°10' W (NAD 27)

**Trevor Channel Entrance Sector
Light** Chart 3671;
position: (SW side of channel entrance):
48°48.74' N, 125°10.82' W (NAD 27)

Trial Islands Light Chart 3424;
position: (S. side of southernmost isl.):
48°23.71' N, 123°18.31' W (NAD 83)

Triangle Island Chart 3625;
position: 50°52' N, 129°05' W (NAD 27)

Tribal Group Chart 3787;
north entrance: 52°03.40' N, 128°18.60' W
(NAD 27)

Tribune Bay Park, Hornby Island
Chart 3527;
anchor Tribune Bay Park:
49°31.32' N 124°37.70' W (NAD 27)

Tribune Channel Chart 3515;
south entrance: 50°40.40' N, 126°14.40' W
(NAD 83)

Tribune Rock Light Chart 3549;
position: (on rock):
50°51.47' N, 127°33.86' W (NAD 83)

Triple Islands Light Chart 3957;
position: (northwesterly rock of the Triple
Islands): 54°17.69' N, 130°52.83' W
(NAD 83)

Triumph Bay Chart 3743;
entrance: 53°29.40' N, 129°43.50' W;
position: 53°27.20' N, 128°41.10' W
(NAD 27)

Trivett Point Light Chart 3740;
position: (northern extremity of Princess
Royal Isl.): 53°18.50' N, 129°02.09' W
(NAD 27)

Trotter Bay Chart 3807;
entrance: 52°53.10' N, 131°52.60' W;
anchor: 52°52.90' N, 131°53.02' W (NAD 27)

Troup Narrows ("Deer Passage")
Charts 3720 (inset), 3940;
position: 52°06.00' N, 128°00.00' W
(NAD 27)

"Troup Narrows Cove"
Charts 3940, 3720;
anchor: 52°16.98' N, 127°59.59' W
(NAD 83)

Trout Bay Charts 3711 (inset), 3734;
public dock: 52°35.44' N, 128°30.92' W;
fuel dock: 52°35.65' N, 128°31.18' W
(NAD 27)

Tsakonu Cove Chart 3545;
anchor: 50°38.37' N, 126°10.74' W (NAD 83)

Tsamspanaknok Bay Chart 3994;
entrance: 54°40.75' N, 130°06.40' W;
anchor: 54°40.00' N, 130°05.87' W (NAD 27)

Tsawwassen Beach
Charts 3499 or 3463;
west end of south breakwater light:
49°00.14' N, 123°07.64' W (NAD 27);
anchor: 49°01.10' N, 123°06.30' W (NAD 27)

Tsawwassen Ferry Breakwater Light
Chart 3499;
position: (W end of S. breakwater):
49°00.14' N, 123°07.64' W (NAD 27)

Tsawwassen Ferry South Light
Chart 3499;
position: (off S. breakwater):
49°00.25' N, 123°07.94' W (NAD 27)

Tsawwassen Light Chart 3499;
position: 49°00.46' N, 123°08.53' W
(NAD 27)

Tsawwassen Range Light Chart 3499;
position: (inner end of terminal):
49°00.40' N, 123°07.68' W (NAD 27)

Tsehum Harbour Chart 3313, p. 7;
entrance light: 48°40.31' N, 123°24.14' W
(NAD 83)

Tsehum Harbour Entrance Light
Chart 3476;
position: 48°40.34' N, 123°24.25' W (NAD 27)

Tsehum Harbour Light Chart 3476;
position: 48°40.31' N, 123°24.14' W
(NAD 27)

Tsibass Lagoon Chart 3547;
entrance: 50°58.44' N, 127°02.60' W
(NAD 83)

Tsowwin Narrows Light Chart 3663;
position: (edge of spit extending out from E.
side of inlet): 49°46.59' N, 126°38.52' W
(NAD 27)

Tsowwin Narrows West Light
Chart 3663;
position: 49°46.41' N, 126°38.63' W
(NAD 27)

Tsum Tsadai Inlet Chart 3717;
entrance: 54°10.45' N, 130°16.43' W;
anchor: 54°10.52' N, 130°14.73' W (NAD 27)

Tsusiat River Chart 3606;
way-point: 48°41.15′ N, 124°56.05′ W
(NAD 27)

Tuck Inlet Chart 3964;
south entrance (Tuck Narrows):
54°23.92′ N, 130°15.35′ W (NAD 27)

Tucker Bay Chart 3512;
anchor: 49°30.12′ N, 124°16.23′ W (NAD 27)

Tugboat Island Light Chart 3475;
position: (on drying rock SE of isl.):
49°08.89′ N, 123°41.03′ W (NAD 27)

Tugboat Passage Chart 3543;
position: 50°24.52′ N, 125°11.88′ W
(NAD 27)

**Tugwell Reef Light and Bell Buoy
D61** Chart 3957;
position: (SE edge of reef):
54°18.68′ N, 130°30.29′ W (NAD 83)

Tuna Point Chart 3544;
anchor: 50°28.54′ N, 125°59.73′ W (NAD 83)

Tuna Point Light Chart 3564;
position: (W. end of pt, E. side of entrance to
Blenkinsop Bay): 50°28.52′ N, 126°00.13′ W
(NAD 83)

Turn Bay Charts 3539, 3543;
anchor: 50°21.12′ N, 125°27.72′ W;
Turn Island light: 50°20.72′ N, 125°26.84′ W
(NAD 83)

Turn Island Light Chart 3543;
position: (S. side of isl.):
50°20.72′ N, 125°27.84′ W (NAD 83)

Turnbull Cove Chart 3547;
anchor: 50°57.45′ N, 126°50.45′ W (NAD 83)

Turnbull Inlet Chart 3784;
entrance: 51°45.25′ N, 128°03.20′ W;
anchor: 51°45.83′ N, 128°02.28′ W (NAD 27)

Turner Bay, Bliss Landing
Chart 3311, 3538;
entrance: 50°02.17′ N, 124°44.18′ W
(NAD 83)

"Turquoise Cove" Chart 3933;
entrance: 55°45.35′ N, 130°08.05′ W
(NAD 27)

Turret Island Chart 3670;
anchor: 48°53.96′ N, 125°20.74′ W (NAD 83)

Tuwartz Inlet Chart 3722 (inset);
entrance: 53°16.70′ N, 129°29.40′ W
(NAD 27)

"Tuwartz Lagoon Cove" Chart 3722;
anchor 3 fathoms (rock pile):
53°19.34′ N, 129°32.72′ W;
anchor 7 fathoms:
53°19.26′ N, 128°32.41′ W (NAD 27)

Tuwartz Narrows Chart 3722;
south entrance: 53°18.38′ N, 129°31.31′ W
(NAD 27)

Tuzo Rock Light Chart 3415;
position: (on rock):
48°25.55′ N, 123°22.42′ W (NAD 27)

Twin Bay Charts 3539, 3538;
entrance: 50°08.80′ N, 125°06.50′ W
(NAD 83)

Twin Islands Light Chart 3495;
position: 49°21.06′ N, 122°53.43′ W
(NAD 83)

Twin Islands, Baker Passage
Charts 3538, 3311;
position Echo Bay:
50°01.22′ N, 124°55.50′ W (NAD 83)

Tyee Point Light Chart 3534;
position: (W. side of entrance to Horseshoe
Bay): 49°22.83′ N, 123°16.32′ W (NAD 83)

Tyee Point Light Chart 3544;
position: (SW pt of W. Thurlow Isl.):
50°23.13' N, 125°47.00' W (NAD 83)

Tyee Spit North Breakwater Light
Chart 3540;
position: 50°02.36' N, 125°14.62' W
(NAD 83)

Tyee Spit Range Light Chart 3540;
position: 50°02.88' N, 125°15.57' W
(NAD 83)

Tyee Spit South Breakwater Light
Chart 3540;
position: 50°02.15' N, 125°14.56' W
(NAD 83)

"Tzartus Cove" Chart 3668, 3671;
anchor: 48°55.14' N, 125°06.25' W (NAD 83)

Tzoonie Narrows Chart 3312;
position: 49°42.45' N, 123°46.64' W
(NAD 27)

Uchucklesit Inlet Chart 3646 (inset);
entrance: 48°59.00' N, 125°00.00' W
(NAD 83)

Ucluelet Inlet Chart 3646
entrance: 48°55.30' N, 125°31.00' W
(NAD 27)

Uganda Passage Charts 3312, p. 19
(inset), 3311, 3518;
position, Uganda Passage light:
50°06.42' N, 125°03.44' W (NAD 27)

Uganda Passage Light Chart 3538;
position: (on Channel Rock between Shark
Spit and Cortes Isl.):
50°05.62' N, 125°02.19' W (NAD 83)

"Underhill Island Coves" Chart 3784;
entrance: 51°46.16' N, 128°03.42' W
(NAD 27)

Union Bay Chart 3527;
position: 49°35.09'N, 124°52.97' W (NAD 27)

Union Passage, Hawkins Narrows
Chart 3722;
north entrance: 53°24.74' N, 129°24.90' W;
anchor: 53°24.90' N, 129°25.24' W (NAD 27)

"Unnamed Cove" Chart 3784;
anchor: 51°38.92' N, 128°04.51' W (NAD 27)

**"Unnamed Cove" Northwest of
Harbour Island** Chart 3663;
entrance: 49°51.5' N, 126°59.5' W (NAD 27)

"Unnamed Cove," Laredo Inlet
Chart 3737;
entrance: 52°43.83' N, 128°46.90' W;
anchor: 52°43.52' N, 128°47.09' W (NAD 27)

"Unnamed Inlet" Chart 3552;
north entrance: 51°02.06' N, 127°19.10' W
(NAD 27)

**Upper Burke Channel (above
Cathedral Point)** Chart 3729;
position ("The Crack"):
52°13.40' N, 127°22.80' W (NAD 27)

Upper Rapids Chart 3537;
position: 50°18.30' N, 125°13.90' W
(NAD 27)

Ursula Channel Charts 3740, 3742;
south entrance: 53°18.60' N, 128°54.00' W
(NAD 27)

Useless Inlet
Charts 3671, 3668; 3646 (inset);
"Useless Bay" anchor:
48°59.45' N, 125°02.12' W;
"Useless Nook" anchor:
48°59.18' N, 125°01.75' W (NAD 27)

Useless Point Light Chart 3956;
position: (on crib, S. side Edye Passage):
54°02.43' N, 130°33.70' W (NAD 83)

Valdes Bay Chart 3664;
anchor SE corner: 49°43.88' N, 126°29.05' W
(NAD 27)

Vananda Cove, Sturt Bay (Marble Bay), Caesar Cove Charts 3536, 3513;
entrance: 49°45.75' N, 124°33.60' W
(NAD 27)

Vancouver Approach Cautionary Light Buoy QA Chart 3481;
position: (W. of Point Grey):
49°16.58' N, 123°19.23' W (NAD 27)

Vancouver Bay Chart 3514;
entrance: 49° 54.70' N, 123°53.30' W
(NAD 27)

"Vancouver Cove," Chart 3920;
anchor north side of two islets:
55° 26.93' N, 129°46.37' W (NAD 83)

Vancouver Harbour, Port of Vancouver Chart 3493;
Lionsgate Bridge south sector light:
49°18.89' N, 123°08.26' W;
Prospect Point red light:
49°18.85' N, 123°08.41' W (NAD 83)

Vancouver Rock Light and Whistle Buoy E54 Chart 3728;
position: (westward of the rock, Milbanke Sound): 52°21.08' N, 128°30.40' W (NAD 27)

Vansittart Point Light Chart 3544;
position: (on pt, S. side of W. Thurlow Isl.):
50°22.61' N, 125°44.60' W (NAD 83)

Varney Bay Chart 3679;
anchor: 50°33.37' N, 127°31.73' W (NAD 83)

Venn Passage and Metlakatla Bay Charts 3955, 3958;
east entrance: 54°18.68' N, 130°22.88' W
(NAD 27);
south entrance (Metlakatla Bay):
54°18.40' N, 130°30.40' W (NAD 83)

Venn Passage Light Chart 3955;
position: (on drying rock):
54°18.73' N, 130°23.48' W (NAD 27)

Vere Cove Chart 3544;
anchor: 50°23.34' N, 125°46.53' W
(NAD 83)

Vernaci Island Light Chart 3664;
position: 49°38.22' N, 126°35.48' W
(NAD 27)

Verney Passage Chart 3743;
south entrance: 53°22.10' N, 129°09.50' W;
north entrance: 53°34.90' N, 128°50.50' W
(NAD 27)

Vesuvius Bay Chart 3313, p.14 ;
public dock in Vesuvius:
48°52.85' N, 123°34.38' W (NAD 83)

Victor Island Light Chart 3664;
position: (N. side of isl.):
49°39.80' N, 126°09.45' W (NAD 27)

Victoria Aero Light Chart 3441;
position: 48°38.73' N, 123°25.20' W
(NAD 27)

Victoria Charts 3641, 3440, 3415;
Odgen Point breakwater light:
48°24.82' N, 123°23.55' W (NAD 27)

Victoria Harbour Cautionary Light Buoy VH Chart 3440;
position: (SW of Brotchie Ledge light):
48°22.53' N, 123°23.48' W (NAD 27)

Victoria Shoal Light Buoy U43 Chart 3442;
position: (NW of Governor Rock):
48°55.20' N, 123°30.93' W (NAD 27)

Vigilance Cove Chart 3551;
entrance: 51°05.30' N, 127°38.40' W;
anchor: 51°05.45' N, 127°38.14' W
(NAD 83)

Village and Indian Channels
Charts 3545, 3546

Village Bay Chart 3539;
anchor north: 50°09.84' N, 125°11.45' W;
anchor south: 50°09.46' N, 125°11.61' W
(NAD 83)

Village Bay, Mayne Island
Chart 3313, p. 9 ;
anchor: 48°50.57' N, 123°19.50' W (NAD 83)

Village Cove Chart 3552;
anchor: 51°10.49' N, 127°24.65' W (NAD 27)

Village Island North Cove Chart 3545;
anchor: 50°37.88' N, 126°33.36' W
(NAD 83)

Village Rocks Light Buoy V20
Chart 3419;
position: (off N. end of rocks): 48°25.98' N,
123°25.54' W (NAD 83)

Viner Sound Chart 3515;
anchor: 50°47.38' N, 126°23.11' W (NAD 83)

Virago Rock Sector Light Chart 3473;
position: (on rock):
49°00.77' N, 123°35.49' W (NAD 27)

Volcanic Cove Chart 3682;
position: 49°58.6' N, 127°13.9' W (NAD 27)

**Von Donop Inlet, Von Donop
Marine Park** Charts 3538, 3312;
anchor: 50°08.55' N, 125°56.63' W (NAD 27)

Waddington Bay Chart 3546;
anchor (west of islet):
50°43.04' N, 126°37.14' W (NAD 83)

Waddington Channel
Charts 3538, 3541;
south entrance: 50°09.80' N, 124°44.30' W
(NAD 27)

Waddington Channel Light
Chart 3541;
position: (Redonda Islands):
50°13.93' N, 124°49.20' W (NAD 83)

Waddington Harbour
Charts 3542, 3312;
position: 50°53.90' N, 124°50.00' W
(NAD 83)

Wahkana Bay Chart 3515;
anchor: 50°49.05' N, 126°17.45' W (NAD 83)

Waiatt Bay Chart 3537;
anchor head of bay:
50°15.83' N, 125°15.24' W (NAD 27)

Wain Rock Light Chart 3441;
position: 48°41.25' N, 123°29.30' W
(NAD 27)

Wakeman Sound Chart 3515;
entrance: 50°56.60' N, 126°29.40' W
(NAD 83)

Wakes Cove, Valdes Island
Chart 3313, p. 19;
anchor: 49°07.56' N, 123°42.21' W (NAD 83)

"Wales Cove" Chart 3994;
entrance: 54°46.75' N, 130°25.70' W;
anchor: 54°47.00' N, 130°25.60' W (NAD 27)

Wales Harbour, "West Cove"
Chart 3960;
entrance: 54°46.60' N, 130°36.80' W;
anchor: 54°45.28' N, 130°35.78' W (NAD 83)

Wales Passage Chart 3994; south
entrance: 54°45.60' N, 130°25.80' W;
north entrance: 54°50.50' N, 130°27.60' W
(NAD 27)

Walkem Islands Chart 3543;
light: 50°21.48' N, 125°31.50' W;
anchor: 50°21.68' N, 125°30.90' W (NAD 27)

Walkem Islands Light Chart 3543;
position: 50°21.47' N, 125°31.42' W
(NAD 83)

Walker (Camp) Island Light
Chart 3787;
position: (S. end of isl., Lama Passage):
52°05.97' N, 128°06.92' W (NAD 27)

"Walker Group Cove" Chart 3574;
anchor: 50°53.94' N, 127°31.88' W (NAD 27)

Walker Rock Light Chart 3442;
position: (Trincomali Channel):
48°55.42' N, 123°29.59' W (NAD 27)

Wallace Bay Charts 3781, 3720, 3729;
anchor: 52°17.50' N, 127°44.95' W
(NAD 27)

Wallace Bight Chart 3734;
entrance: 52°43.80' N, 128°25.90' W;
anchor: 52°43.64' N, 128°23.90' W (NAD 27)

Wallace Islands Chart 3548;
entrance: 50°57.44' N, 127°27.57' W;
anchor: 50°57.56' N, 127°27.37' W (NAD 83)

Walsh Cove Marine Park Charts 3541;
position: 50°16.15' N, 124°48.00' W
(NAD 27)

Walsh Rock Chart 3737;
anchor position: 52°38.20' N, 128°47.67' W
(NAD 27)

Walsh Rock Light Chart 3742;
position: 52°38.23' N, 128°57.25' W
(NAD 27)

Walter Bay Chart 3313, p. 10;
anchor: 48°50.67' N, 123°29.16' W (NAD 83)

Walters Cove Chart 3651
entrance: 50° 01.30' N, 127°22.06' W
anchor: 50° 01.75' N, 127°22.37' W
(NAD 83)

Ward Channel Chart 3784;
south entrance: 51°45.38' N, 128°03.47' W;
north entrance: 51°47.00' N, 128°02.60' W
(NAD 27)

Warn Bay Chart 3649;
anchor: 49°15.45' N, 125°44.70' W (NAD 27)

Warner Bay Chart 3552;
entrance: 51°02.78' N, 127°05.50' W
(NAD 27)

Warren Islands, Call Inlet
Charts 3564, 3545;
entrance: 50°34.40' N, 126°11.60' W (NAD 83)

Watcher Island Light Chart 3934;
position: (W. pt of isl.):
51°16.28' N, 127°43.04' W (NAD 83)

Water Bay Chart 3535
(Welcome Passage inset);
anchor: 49°29.57' N, 123°58.77' W (NAD 27)

Waterfall Inlet Chart 3921;
south entrance: 51°36.95' N, 127°43.57' W
(NAD 83)

Watson Bay Chart 3734;
entrance: 52°41.20' N, 128°25.95' W (NAD 27)

Watson Cove Chart 3515;
anchor: 50°50.94' N, 126°18.33' W (NAD 83)

Watson Rock Light Chart 3773;
position: (on the rock, western entrance to
Grenville Ch.): 53°55.42' N, 130°10.27' W
(NAD 27)

Watt Bay Chart 3784;
southwest entrance:
51°50.32' N, 128°08.23' W (NAD 27)

Watts Narrows Chart 3772 (inset);
west entrance: 53°48.65' N, 129°57.23' W;
east entrance: 53°49.00' N, 129°56.85' W
(NAD 27)

Waump Creek Chart 3552;
position; 51°11.15' N, 126°55.15' W (NAD 27)

Wawatle Bay Chart 3552;
narrows: 51°02.06' N, 127°18.04' W;
anchor: 51°02.04' N, 127°16.65' W (NAD 27)

Wearing Point Light Chart 3781;
position: (on pt, Cousins Inlet):
52°18.02' N, 127°45.59' W (NAD 27)

Wedge Island Light Chart 3546;
position: (on isl., entrance to Knight Inlet):
50°38.08' N, 126°43.27' W (NAD 83)

Wedge Point Sector Light Chart 3711;
position: 52°36.64' N, 128°31.01' W (NAD 27)

Weeteeam Bay Chart 3710 (inset);
entrance: 52°29.70' N, 129°02.20' W
(NAD 27)

Weewanie Hot Springs Chart 3743;
cove position: 53°41.78' N, 128°47.30' W
(NAD 27)

Wehlis Bay Chart 3547;
west entrance: 50°50.20' N, 126°54.50' W
(NAD 83)

Weinburg Inlet, Dunn Passage
Chart 3719 (inset);
entrance: 53°06.87' N, 129°32.62' W;
anchor (Weinburg Inlet):
53° 07.52' W, 129°30.86' W (NAD 27)

Welbury Bay Chart 3313, p. 10;
anchor: 48°50.94' N, 123°26.21' W (NAD 83)

Welcome Harbour Chart 3909 (inset);
north entrance: 54°02.00' N, 130°37.45' W;
anchor: 54°00.14' N, 130°39.67' W (NAD 83)

Weld Cove Chart 3737;
entrance (between Kohl and Pocock islands):
52°48.15' N, 128°45.10' W;
anchor: 52°49.10' N, 128°46.00' W (NAD 27)

Wellbore Channel Chart 3544;
south entrance: 50°25.50' N, 125°43.50' W
(NAD 83)

"Wellbore Cove" Chart 3544;
position: 50°27.10' N, 125°46.00' W
(NAD 83)

Wells Passage Chart 3547;
east entrance: 50°53.00' N, 126°52.75' W
(NAD 83)

**Wentworth Rock Light and Whistle
Buoy N31** Chart 3549;
position: 50°57.38' N, 127°29.97' W
(NAD 83)

West Bay, Ruxton Island
Chart 3313, p. 18;
position: 49°04.70' N, 123°42.48' W
(NAD 83)

"West Blenkinsop Bay"
Charts 3564, 3544;
anchor: 50°28.72' N, 126°01.70' W (NAD 83)

"West Fleming Cove" Chart 3668, 3671;
anchor: 48°53.05' N, 125°08.68' W (NAD 83)

**"West Home Bay" (South Shore
Rivers Inlet)** Chart 3934;
anchor: 51°23.88' N, 127°43.26' W (NAD 83)

West Narrows, Skidegate Channel
Chart 3891 (inset);
east entrance: 53°08.92' N, 132°20.30' W;
west entrance: 53°08.76' N, 132°22.10' W
(NAD 83)

West Redonda Island Light
Chart 3541;
position: 50°14.58' N, 124°58.60' W
(NAD 83)

"West Whitepine Cove" Chart 3648;
inner cove anchor:
49°17.88' N, 125°58.30' W (NAD 27)

Westerly Light Chart 3491;
position: 49°15.45′ N, 123°16.70′ W
(NAD 83)

Westerman Bay Chart 3552;
entrance: 51°08.25′ N, 127°28.00′ W;
anchor: 51°09.18′ N, 127°27.17′ W (NAD 27)

Westview Boat Harbour North Light
Chart 3536;
position: (on breakwater):
49°50.37′ N, 124°31.85′ W (NAD 27)

**Westview Fishing Harbour North
Light** Chart 3536;
position: (S. extremity of northerly breakwater): 49°50.05′ N, 124°31.70′ W (NAD 27)

**Westview Fishing Harbour South
Light** Chart 3536;
position: (seaward end of southerly breakwater): 49°49.90′ N, 124°31.68′ W (NAD 27)

Weynton Passage Chart 3546;
south entrance: 50°34.10′ N, 126°48.30′ W
(NAD 83)

Whale Channel Chart 3742;
SW entrance: 53°04.20′ N, 129°15.00′ W;
northeast
entrance: 53°17.10′ N, 129°07.90′ W
(NAD 27)

**Whale Rock Bifurcation Light Buoy
VC** Chart 3419;
position: 48°26.60′ N, 123°26.66′ W
(NAD 83)

Whaleboat Island Marine Park
Chart 3313, p. 18;
position: 49°04.50′ N, 123°41.60′ W
(NAD 83)

Whaletown Bay Charts 3311, 3538;
public float: 50°06.50′ N, 125°03.05′ W
(NAD 83)

Whaletown Bay Entrance Light
Chart 3538;
position: (Cortes Isl.):
50°06.42′ N, 125°03.44′ W (NAD 83)

Wheeler Island Light Chart 3746;
position: (SW corner of isl.):
53°33.00′ N, 130°08.58′ W (NAD 27)

Whelakis Lagoon Chart 3552;
entrance: 50°58.23′ N, 127°11.73′ W;
anchor: 50°57.90′ N, 127°12.18′ W (NAD 27)

Whiffin Spit Light Chart 3430;
position: (E. end of spit, Sooke Inlet):
48°21.52′ N, 123°42.61′ W (NAD unknown)

Whirlpool Rapids Chart 3544;
Carterer Point light:
50°27.53′ N, 125°45.83′ W (NAD 83)

Whirlwind Bay Chart 3785 (inset), 3784;
position: 51°51.78′ N, 127°51.73′ W
(NAD 27)

Whiskey Bay Chart 3730;
position: 52°21.74′ N, 126°52.52′ W
(NAD 27)

Whiskey Bay Chart 3933;
anchor: 55°01.98′ N, 130°10.09′ W (NAD 27)

Whisky Cove Charts 3785, 3787;
position: 52°09.45′ N, 128°06.15′ W
(NAD 27)

White Islets Light Chart 3311;
position: (E. of Sechelt):
49°25.10′ N, 123°42.63′ W (NAD 83)

White Point Light Chart 3785;
position: (on pt): 52°04.52′ N, 127°57.85′ W
(NAD 27)

White Rock Breakwater Light
Chart 3463;
position: 49°01.00′ N, 122°48.40′ W (NAD 27)

White Rock Chart 3463;
breakwater light: 49°01.00' N, 122°48.40' W
(NAD 27)

Whitepine Cove Chart 3648;
entrance: 49°18' N, 125°57' W (NAD 27)

Whiterock Passage 1 Range Light
Chart 3537;
position: (W. side of Read Isl.):
50°14.77' N, 125°06.14' W (NAD 27)

Whiterock Passage 2 Range Light
Chart 3537;
position: (W. side of Read Isl.):
50°14.64' N, 125°06.37' W (NAD 27)

Whiterock Passage Chart 3537 (inset);
south range entrance:
50°14.58' N, 125°06.78' N;
north range entrance:
50°15.12' N, 125°05.90' W (NAD 27)

Whiterock Passage Light Chart 3537;
position: (SW end of passage):
50°14.58' N, 125°06.63' W (NAD 27)

Whitesand Cove Chart 3648;
anchor: 49°15.04' N, 126°04.60' W (NAD 27)

Whitesand Island Light Chart 3959;
position: 54°30.77' N, 130°44.77' W
(NAD 83)

Wiah Point Light and Bell Buoy C50
Chart 3892;
position: (N. of pt):
54°07.37' N, 132°18.65' W (NAD 27)

Wiah Point Light Chart 3892;
position: (outer end of reefs, NE of pt):
54°06.98' N, 132°18.48' W (NAD 27)

**Wickaninnish Bay Cautionary Light
Buoy** Chart 3685;
position: 49°02.20' N, 125°48.00' W
(NAD 83)

"Wide Awake Cove" Chart 3787, 3786;
entrance: 52°03.00' N, 128°14.16' W;
anchor: 52°02.76' N, 128°14.38' W
(NAD 27)

Wigham Cove Charts 3940, 3720;
entrance: 52°16.50' N, 128°10.60' W;
anchor (east end):
52°16.94' N, 128°10.00' W (NAD 83)

Wilbraham Point Light Chart 3535;
position: (SW side of Grant Isl., off pt):
49°30.68' N, 123°58.15' W (NAD 27)

Wilby Shoals Light Buoy P60
Chart 3539;
position: (S. of shoals):
49°58.97' N, 125°09.08' W (NAD 83)

"Wilcox Group Cove"
Charts 3761, 3927;
anchor: 53°55.17' N, 130°39.88' W
(NAD 27)

Wilfred Point Light Chart 3539;
position: (on pt): 50°07.84' N, 125°21.54' W
(NAD 83)

"Wilkie Point Cove" Chart 3550;
north entrance: 51°08.30' N, 127°44.00' W;
south entrance point:
51°08.15' N, 127°43.70' W;
anchor: 51°08.44' N, 127°43.47' W
(NAD 83)

Will Rock Light Chart 3685;
position: (on rock):
49°08.23' N, 125°58.53' W (NAD 83)

William Head Light Chart 3461;
position: (on extremity of head):
48°20.58' N, 123°31.59' W (NAD 27)

Williamson Passage Light Chart 3664;
position: 49°39.18' N, 126°22.30' W
(NAD 27)

Willis Bay Charts 3761, 3747, 3927;
entrance: 53°47.00′ N, 130°32.20′ W;
mooring buoys: 53°48.31′ N, 130°32.27′ W
(NAD 27)

Wilson Bay Chart 3934;
entrance: 51°27.55′ N, 127°39.05′ W (NAD 83)

Wilson Inlet Chart 3746;
entrance: 53°33.45′ N, 129°56.80′ W (NAD 27)

**Wilson Rock Light and Bell Buoy
E75** Chart 3726;
position: 52°40.00′ N, 128°57.92′ W
(NAD unknown)

Windsor Cove Chart 3729;
position (head of cove):
51°56 31′ N, 127°52.96′ W (NAD 27)

Windy Bay Chart 3685;
anchor: 49°08.31′ N, 125°49.00′ W (NAD 83)

Windy Bay Chart 3730;
position: 52°20.70′ N, 126° 54.90′ W (NAD 27)

Windy Bay Chart 3962;
entrance: 52°47.30′ N, 128°13.60′ W;
anchor: 52°47.07′ N, 128°12.60′ W (NAD 27)

Wing Dam Range Light Chart 3490;
position: (close behind Steveston Jetty):
49°07.97′ N, 123°14.00′ W (NAD 83)

Winter Cove Marine Park
Chart 3313, p. 12;
anchor: 48°48.68′ N, 123°11.57′ W
(NAD 83)

Winter Harbour Chart 3686;
public dock: 50°30.77′ N, 128°01.73′ W
(NAD 83)

Winter Inlet Chart 3994;
entrance: 54°50.50′ N, 130°27.60′ W;
anchor: 54°48.30′ N, 130°25.83′ W (NAD 27)

"Withered Point Coves," Lina Island
Chart 3981;
anchor (east of spit):
53°13.22′ N, 132°08.32′ W;
anchor (west of spit):
53°13.33′ N, 132°08.88′ W (NAD 83)

Wizard Islet Light Chart 3671;
position: (on the islet):
48°51.49′ N, 125°09.52′ W (NAD 27)

Wood Bay Chart 3311;
position: 49°32.94′ N, 123°59.44′ W
(NAD 83)

Woodlands Light Chart 3495;
position: (N. side of Indian Arm):
49°20.47′ N, 122°55.17′ W (NAD 83)

Woods Bay Chart 3547;
anchor: 50°56.10′ N, 126°51.14′ W
(NAD 83)

Woods Lagoon Chart 3552;
entrance: 51°00.90′ N, 127°17.52′ W
(NAD 27)

Woods Nose and "Mill Cove"
Chart 3647;
anchor: 48°32.87′ N, 124°26.28′ W (NAD 27)

Woodward Island Light Chart 3490;
position: (NE end of isl.):
49°06.39′ N, 123°07.49′ W (NAD 83)

**Woodward Island Upstream Range
Light** Chart 3490;
position: 49°06.38′ N, 123°08.24′ W
(NAD 83)

Wootton Bay Charts 3559, 3312, 3538;
anchor: 50°04.93′ N, 124°43.25′ W (NAD 27)

Work Bay Chart 3738;
entrance: 52°45. 95′ N, 128°28.90′ W;
anchor: 52°47.05′ N, 128°28.80′ W (NAD 27)

Work Channel Chart 3963;
entrance: 54°39.00' N, 130°26.70' W
(NAD 83)

Work Island Light Chart 3740;
position: (W. end of isl.):
53°10.72' N, 128°41.55' W (NAD 27)

Worsfold Bay Chart 3963;
position: 54°34.10' N, 130°17.85' W
(NAD 83)

Wright Inlet Chart 3746 (inset);
position (Wright Narrows):
53°31.04' N, 129°52.02' W (NAD 27)

Wright Sound Chart 3742;
position: 53°20.50' N, 128°13.00' W
(NAD 27)

Wyclees Lagoon Chart 3931;
entrance: 51°17.40' N, 127°20.84' W
(NAD 83)

Yaculta Chart 3540;
public float: 50°01.39' N, 125°11.70' W
(NAD 83)

Yeatman Bay Chart 3537;
anchor: 50°14.01' N, 125°10.78' W (NAD 27)

"Yellow Bluff Bight" Chart 3663;
anchor: 49°51.47' N, 127°06.90' W
(NAD 27)

Yellow Bluff Sector Light Chart 3546;
position: (on shoal):
50°35.22' N, 126°57.05' W (NAD 83)

Yeo Cove Chart 3940;
anchor: 52°17.78' N, 128°10.98' W
(NAD 83)

York Point Light Chart 3742;
position: (southern pt of peninsula, Gil Isl.):
53°05.48' N, 129°10.38' W (NAD 27)

Yorke Island, Nichols Bay Chart 3544;
position: 50°26.18' N, 125°58.05' W (NAD 83)

Young Bay Chart 3648;
anchor: 49°25.68' N, 126°13.06' W (NAD 27)

Yuculta Rapids and Dent Rapids
Chart 3543 (inset);
light, northeast corner Gillard Islands:
50°23.49' N, 125°09.26' W;
light, Little Dent Island:
50°24.53' N, 125°12.52' W (NAD 27)

Zeballos Inlet Chart 3663;
public dock: 49°58.72' N, 126°50.67' W
(NAD 27)

Zeballos Inlet Light Chart 3663;
position: (on small isl., W side of inlet):
49°56.78' N, 126°48.90' W (NAD 27)

Zeballos Inlet North Light Chart 3663;
position: (on pt): 49°57.58' N, 126°50.68' W
(NAD 27)

Zeballos Inlet South Light Chart 3663;
position: (on pt on W shore):
49°54.45' N, 126°47.97' W (NAD 27)

Zero Rock Light Chart 3440;
position: 48°31.43' N, 123°17.43' W
(NAD 27)

Zuclarte Channel Light Chart 3664;
position: (E. side of Bligh Isl.):
49°39.15' N, 126°28.97' W (NAD 27)

Zuclarte Channel South Light
Chart 3664;
position: (S. entrance to channel):
49°35.75' N, 126°31.20' W (NAD 27)

Zumtela Bay Chart 3963;
entrance: 54°34.85' N, 130°21.75' W;
anchor: 54°34.95' N, 130°22.37' W
(NAD 83)

Personal waypoints or corrections from website:
www.fineedge.com

Personal waypoints or corrections from website: www.fineedge.com

Personal waypoints or corrections from website: www.fineedge.com

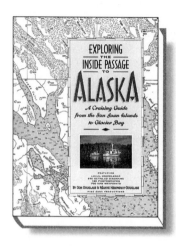

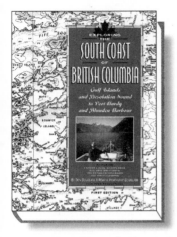
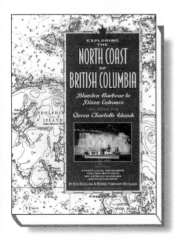